Inorganic Chemistry Concepts
Volume 11

Shinichi Kawaguchi

Variety in Coordination Modes of Ligands in Metal Complexes

With 56 Figures

Springer-Verlag
Berlin Heidelberg NewYork London Paris Tokyo

Professor Dr.
Shinichi Kawaguchi

Kinki University,
School of Medicine
Premedical Building,
377-2 Ohnohigashi, Osakasayama,
Osaka 589, Japan

ISBN 978-3-642-50150-0   ISBN 978-3-642-50148-7 (eBook)
DOI 10.1007/978-3-642-50148-7

Library of Congress Cataloging-in-Publication Data
Kawaguchi, Shinichi, 1920–Variety in coordination modes of ligands in metal com-
plexes. (Inorganic chemistry concepts) Bibliography: p. Includes index. 1. Complex
compounds. 2. Chemical bonds. 3. Ligands. I. Title. II. Series.
QD474.K39 1987 541.2'242   87–23550
ISBN 978-3-642-50150-0

© Springer-Verlag, Berlin Heidelberg 1988
Softcover reprint of the hardcover 1st edition 1988

Typesetting: Hagedorn, Berlin;

2152/3020-543210

Preface

In 1893, Alfred Werner proposed the coordination theory and demonstrated that stereochemistry is not limited to carbon compounds, but is a general phenomenon encompassing metal coordination compounds. This epoch-making theory not only succeeded in systematizing the transition-metal complexes but also afforded the foundation of modern structural inorganic chemistry.

Since then, the number of synthetic metal complexes has rapidly increased. Coordination chemistry has been regarded as a special field of inorganic chemistry although an increasing number of organic molecules and ions are involved as ligands in metal complexes.

The rapid growth of organometallic chemistry during the last four decades widened the scope of coordination chemistry tremendously. Many nonclassically bonded compounds have joined the field as new members; these show metal-to-carbon bonding that cannot be explained by the simple two center—two electron (2c–2e) covalent bond. Transition metal complexes of alkenes, alkynes, arenes, and other unsaturated organic molecules involving the interactions of π electrons with metal orbitals, constitute the most important class of these nonclassical or non-Werner-type complexes. Furthermore, organometallic compounds brought about new types of reactions such as oxidative addition and migratory insertion which had not been important for classical coordination compounds.

In recent years, metal complexes are gaining importance as catalysts in synthetic and enzyme-like reactions. For example, nitrogen fixation, C_1 chemistry, and asymmetric syntheses are among the most important topics in chemistry. It should be noticed that the reactivity of metal complexes depends not only on the nature of metal and substrate ligands but also on their interactions with other coexisting ligands.

Here it will be shown that ligands can exert various modes of coordination in metal complexes. The bonding mode is a probe reflecting the chemical environment of the particular ligand. It is influenced by the nature of the central metal atom and of ancillary ligands, and also by solvents and packing effects in the solid state. When a substrate is bonded to a metal atom, its behavior as a ligand may be related to the activation of a reaction.

The hydride anion will first serve as an example of the most simple monoatomic ligands; it plays an important role in hydrogenation reactions. Carbon monoxide and dinitrogen are diatomic ligands which are receiving much attention at present. The thiocyanate anion is a suitable ambidentate ligand which exemplifies the phenomenon of linkage isomerism. Lastly, β-dicarbonyl compounds will be treated as representative polyatomic ligands. They constitute a class of most important ligands which usually serve as monoanionic chelating agents in classical metal complexes. Very recently, they have been disclosed to exhibit various novel modes of coordination not only as monoanions but also as di- and tri-anions and as uncharged molecules.

It is a pleasure for me to thank Professor Hideo Yamatera of Daido Institute of Technology, Nagoya, who encouraged me to write this book, and Mrs. Yu Isobe of Nara Women's University for her kind preparation of figures. I am also grateful to the following organizations for permission of using materials from several journals: The American Chemical Society, Elsevier Sequoia S.A., Pergamon Journals Ltd., Plenum Publishing Corporation, The Royal Society of Chemistry, and VCH Verlagsgesellschaft mbH.

Osaka, September 1987   Shinichi Kawaguchi

Contents

1	**Introduction**	1
1.1	Classification of Ligands	1
1.2	Linkage Isomerism	3
2	**Monoatomic Ligands**	7
2.1	Coordination Modes for the Hydride Ligand........	7
2.1.1	Terminal Hydride Ligands	8
2.1.1.1	Preparative Methods	8
2.1.1.2	Characterization	8
2.1.1.3	Tricapped Trigonal Prism Structure	9
2.1.1.4	Octahedral Structure	10
2.1.1.5	Pentagonal-Bipyramid Structure	10
2.1.1.6	*Trans* Influence of the Hydride Ligand	11
2.1.2	Unsupported M—H—M Linkage	12
2.1.3	Dinuclear $M(\mu\text{-H})_nM$ and Mixed-bridged Systems	14
2.1.3.1	Preparative Methods	14
2.1.3.2	Characterization	14
2.1.3.3	$M(\mu\text{-H})_2M$ Systems	15
2.1.3.4	$M(\mu\text{-H})_3M$ Systems	16
2.1.3.5	$M(\mu\text{-H})_4M$ Systems	17
2.1.3.6	$M(\mu\text{-H})(\mu\text{-X})_n$ Systems	18
2.1.4	Edge-bridging (μ_2) and Face-bridging (μ_3) Hydride Ligands in Metal Clusters	18
2.1.5	Systems with an Interstitial Hydrogen Atom	20
2.1.6	Metal Clusters Including an Interstitial Light Atom other than Hydrogen	21
2.2	Chemical Reactions of Hydride Ligands	21
2.2.1	Reactions with Acids	23
2.2.2	Reactions with Bases	23
2.2.3	Reactions with Halogens and Organic Halides	23
2.2.4	Intramolecular Migratory Insertion	24
2.2.5	Reductive Elimination Reactions	24

2.3   Role of Rhodium Hydride Complexes in the
          Catalytic Hydrogenation of Olefins   25
2.3.1   Homogeneous Activation of Dihydrogen by Metal
          Complexes in Solution   25
2.3.1.1   Three Modes of H_2 Activation   25
2.3.1.2   Dihydride Formation by Oxidative Addition of H_2 ...   27
2.3.1.3   Dihydrogen Complex   27
2.3.2   Reaction Pathway for the Hydrogenation of Olefins
          Catalyzed by RhCl(PPh$_3$)$_3$ 29
2.3.3   Hydrogenation of Olefins Catalyzed by Cationic
          Rh(I) Complexes, [Rh(PP)(S)$_2$]$^+$ 31
2.3.3.1   Formation of Cationic Rh(I) Complexes   31
2.3.3.2   Mechanism of Olefin Hydrogenation   31
2.3.3.3   Asymmetric Hydrogenation......................   33

3   Diatomic Ligands   35
3.1   Coordination Modes for Carbon Monoxide   35
3.1.1   Terminal Bonding and μ(C) Bridging   35
3.1.2   M$_x$—CO—M$'$ Bridging   36
3.1.2.1   M—CO—M$'$ Systems   37
3.1.2.2   M$_2$—CO—M$'$ Systems......................   37
3.1.2.3   M$_3$—CO—M$'$ Systems......................   38
3.1.3   CO Bridge Involving the η^2 (side-on) Linkage   39
3.1.3.1 $\mu(\eta^1, \eta^2)$ Mode.........................   40
3.1.3.2 $\mu_3(\eta^1, 2\eta^2)$ Mode.......................   41
3.1.3.3 $\mu_4(3\eta^1, \eta^2)$ Mode.......................   41
3.2   CO Cleavage and Reduction   43
3.3   Coordination Modes for Dinitrogen   45
3.3.1   End-on Unidentate Coordination.................   47
3.3.2   End-on Bridging   49
3.3.3   End-on μ_3-Bridging   50
3.3.4   Side-on Coordination   52
3.3.5   Side-on Bridging   52
3.3.6   End-on:side-on μ_3-Bridging   54
3.4   Protonation of the Coordinated Dinitrogen   55

4   Triatomic Ligands   59
4.1   Coordination Modes for the Thiocyanate Ion   59
4.1.1   One-end Bridging   59
4.1.1.1 μ(N) Bridging   60
4.1.1.2 μ(S) Bridging   61
4.1.1.3 μ_3(S) Bridging   62
4.1.2   End-to-end Bridging...........................   62

4.1.2.1   Single μ(S,N) Bridging 62
4.1.2.2   Double μ(S,N) Bridging 63
4.1.2.3 μ_3(2S,N) Bridging 64
4.1.2.4 μ_4(3S,N) Bridging 64
4.2   Infrared Spectroscopy for Determining the
          Coordination Modes of the Thiocyanate Ligand 65
4.3   Factors Influencing the Relative Stabilities of the
          N-bonded and S-bonded Thiocyanate Complexes 67
4.3.1   Principle of Hard and Soft Acids and Bases (HSAB) .. 67
4.3.2   Electronic Effects of Ancillary Ligands 69
4.3.3   Steric Effects of Ancillary Ligands 70
4.3.4   Solvent and Counterion Effects..................... 72

5   **Polyatomic Ligands: β-Dicarbonyl Compounds** 75
5.1   Three Coordination Modes for Neutral Molecules 75
5.1.1   Keto-enol Tautomerism and Structures of
          Enol Molecules 75
5.1.2   Metal Complexes of Neutral Molecules 79
5.1.3   O,O'-Chelation of the Keto Molecules 81
5.1.4   O-Unidentate Coordination of Enol 82
5.1.5 η^2(C,C') Coordination of Enol 83
5.2   Coordination Modes for Monoanions 84
5.2.1   O,O'-Chelation and Bridging 84
5.2.2   Central Carbon Bonding and C,O,O -Bridging 87
5.2.3   Outer-Sphere Coordination 91
5.2.4   O-Unidentate Linkage 92
5.2.5 η-Allylic Coordination 96
5.2.6   Terminal Carbon Bonding.......................... 97
5.3   Coordination Modes for Dianions 98
5.3.1   Central Carbon Bonding 98
5.3.2   Chelation through Terminal Carbons 98
5.3.3   Dienediolate Chelation............................. 98
5.3.4   C,O,O'-Bridging 99
5.3.5 η-Allylic Coordination 100
5.3.6   C,O-Chelation 102
5.3.7 η^3,O,O'-Bridging 104
5.4 η^3,C,O-Bridging of the Acetylacetonate Trianion...... 105
5.5   Concluding Remarks 106

6   **Abbreviations** 107

7   **References** 109

Subject Index 121

1 Introduction

1.1 Classification of Ligands[1]

A coordination compound or a metal complex is composed of a central metal atom or ion and several ligands. A ligand has at least one pair of electrons which can be donated to the metal in forming the coordination linkage. Thus, ligands may also be called Lewis bases, and metal atoms with incomplete valence electron shells are Lewis acids.

There are various ways of classifying ligands. Primarily, they are classified on the basis of their electric charge. Ammonia, H_2O, CO, PR_3, pyridine, and many kinds of organic molecules serve as neutral ligands. On the other hand, Cl^-, CO_3^{2-}, and PO_4^{3-} are examples of mono-, di-, and tri-anionic ligands, respectively. The ethylenediamine-tetraacetate anion, $[(OOC)_2NCH_2CH_2N(COO)_2]^{4-}$ (EDTA) is a tetra-anionic ligand. Even cations, such as $NH_2NH_3^+$ and $NH_2CH_2CH_2NH_3^+$ can coordinate to a metal ion since they preserve an unshared electron pair.

Ligands may also be classified by the number of atoms composing them. Thus, H^-, N_2, N_3^-, NO_3^-, and SO_4^{2-} are mono-, di-, tri-, tetra-, and penta-atomic ligands, respectively, and there are many known polyatomic ligands.

The ligands in $[Co(NH_3)_6]^{3+}$, $Ni(CO)_4$, and $[PtCl_4]^{2-}$ are bound to a single metal atom and said to be unidentate. Ligands having more than one donor atom may become bound to a metal atom via two donor atoms, thus forming a chelate ring, as in:

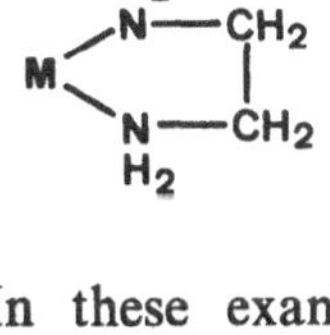

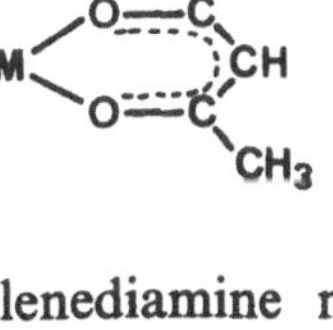

In these examples, the ethylenediamine molecule (en) and the acetylacetonate anion (acac) are bidentate. Similarly, tridentate, tetradentate, and polidentate ligands are known. Thus, the EDTA anion usually functions as a hexadentate ligand, occupying six coordination sites of a metal ion.

When a ligand connects two or more metal atoms, it is called a bridging ligand. For example, two Cl^- ions and three CO molecules in:

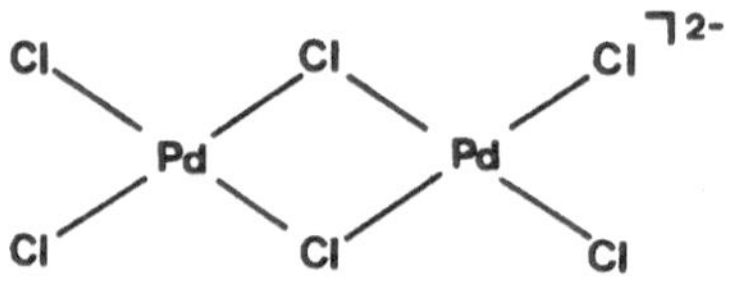

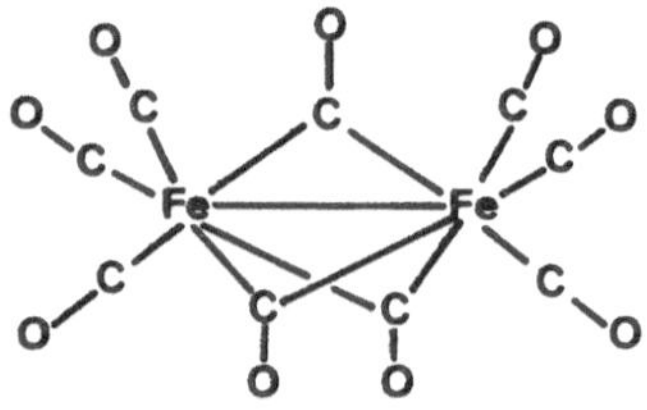

respectively, are bridging; the coordination mode is indicated by the symbol μ_2 or μ; $[Pd_2(\mu\text{-}Cl)_2Cl_4]^{2-}$ and $Fe_2(\mu\text{-}CO)_3(CO)_6$.

The most important distinctions between ligands are based on the electronic properties. Many ligands act as simple electron-pair donors, forming complexes with all types of Lewis acids: metal ions and molecules such as BF_3 and $AlCl_3$. Some other ligands serve not only as σ-donors (σ-base) but also as π-acceptors (π-acid). They usually form compounds with transition metal atoms, since this kind of interaction occurs owing to the special properties of both metal and ligand. The transition metal has d orbitals which can be used in bonding, and the ligand must have vacant orbitals suitable for π-bonding. For example, the phosphorus atom in a tertiary phosphine, $:PR_3$, has vacant $3d$ orbitals of low energy. In the case of the nitrogen atom in ammonia or amines, $:NR_3$, on the other hand, the lowest energy d orbitals are too high to be utilized.

The most important π-acceptor ligand is carbon monoxide. It has no measurable basicity to a proton and gives no stable complexes with strong Lewis acids such as BF_3 and $AlCl_3$. However, CO forms quite stable complexes with soft metal atoms (p. 67) in low oxidation states, giving rise to numerous metal carbonyls and related compounds. The bonding mechanism of CO is synergic: the drift of metal electrons into CO orbitals, "π-bonding", enhances its basicity, strengthening the "σ-bonding". Besides CO there are various ligands for which π-bonding is important, e.g. isocyanides, substituted phosphines, arsines or sulfides, dinitrogen and nitric oxide.

Alkenes, alkynes, and aromatic ring systems such as $C_5H_5^-$, C_6H_6, $C_7H_7^+$ and $C_8H_8^{2-}$ also form π-complexes. They donate the π-electrons to a metal atom (σ-bonding) and accept the electron density from filled metal orbitals into antibonding orbitals on the carbon atoms (π-bonding) in the synergic fashion. Zeise's salt $K[Pt(C_2H_4)Cl_3]$ and ferrocene $Fe(\eta\text{-}C_5H_5)_2$ are the best-known examples. The symbol η is used to signify that all carbon atoms of the ring are bonded to the metal atom. When it is necessary to indicate unambiguously the number of carbon atoms bonded, it is given as a superscript to η. For example, the formula $Fe(\eta^1\text{-}C_5H_5)(\eta^5\text{-}C_5H_5)(CO)_2$ for dicarbonyl(*monohapto*-cyclopentadienyl)(*pentahapto*-cyclopentadienyl)iron(II) shows that one C_5 ring is bonded to iron via one single carbon atom, while the other ring is bonded via all five carbons.

1.2 Linkage Isomerism[2-6]

Since the nitrogen atom in ammonia, organic amines, pyridine and its derivatives has only one pair of unshared electrons and can coordinate with four other atoms at most, these molecules always serve as a unidentate ligand; however, they are rather exceptional. Even monoatomic anions such as Cl^-, O^{2-}, and N^{3-} carry more than one pair of unshared electrons and can act not only as unidentate, but also as bridging ligands connecting two or more metal atoms.

With an increasing number of donor atoms involved, the possible coordination modes of a ligand increase markedly. In the succeeding chapters, representative examples of mono-, di-, tri-, and poly-atomic ligands will be described along with their coordination modes.

When two metal complexes have identical compositions and a common ligand adopts different coordination modes in them, these complexes are called *linkage isomers*. The first examples were $[Co(NH_3)_5(NO_2)]^{2+}$ and $[Co(NH_3)_5(ONO)]^{2+}$ which were prepared by S. M. Jørgensen in 1894.[2,6]

Jørgensen deduced that the nitrite anion is bonded to cobalt with N in the former "nitro" complex and with O in the latter "nitrito" complex, since they are yellow and pink analogously to $[Co(NH_3)_6]^{3+}$ and $[Co(NH_3)_5(H_2O)]^{3+}$, respectively.

Both the kinetic[7,8] and ^{18}O tracer[9] studies showed that production of the nitrito complex from $[Co(NH_3)_5OH]^{2+}$ is not a simple ligand substitution process, but is a nitrosation which takes place in a buffered HNO_2/NO_2^- solution without Co-O bond cleavage. The proposed mechanism is shown by the following equations:

$$2HNO_2 \rightleftarrows N_2O_3 + H_2O$$
$$[Co(NH_3)_5OH]^{2+} + N_2O_3 \rightarrow (NH_3)_5Co\text{---}O \cdots H^{\rceil 2+}$$
$$\vdots$$
$$O\text{---}N \cdots ONO$$
$$\rightarrow [(NH_3)_5CoONO]^{2+} + HNO_2$$

This conclusion was later unambiguously confirmed by X-ray crystallography.[10]

Vibrational spectroscopy is a powerful tool of deducing the mode of nitrite coordination in metal complexes.[11] The nitro complex usually exhibits the N—O stretching absorptions in the 1300–1340 cm^{-1} (v_s) and 1360–1430 cm^{-1} (v_{as}) regions, while the v(N—O) and v(N=O) bands from the nitrito complex appear in the 1050–1110 cm^{-1} and 1400–1485 cm^{-1} regions, respectively. In addition, nitrito complexes lack the wagging modes near 620 cm^{-1} which appear in all nitrito complexes.

The O-bonded nitrito complex is thermodynamically less stable at room temperature and usually prepared at lower temperatures. Upon heating both in solution and in the solid state,[10] it readily isomerizes to the stable N-bonded nitro form by an intramolecular process;[12] this was clarified by the ^{18}O tracer experiments in solution,[9] and supported by high-pressure studies.[13] At present, X-ray structures of a great number of metal complexes containing the N-bonded or O-bonded nitrite ligand are known.[6]

The second example of linkage isomerism was found by Basolo et al. in 1963 for the thiocyanate complexes.[14] The reactions of [Pd(SCN)$_4$]$^{2-}$ with triphenylarsine in ethanol at 0 °C and with bipyridine in the same solvent at −78 °C gave the S-bonded complexes, which on heating isomerized to the N-bonded form:

$$[Pd(SCN)_4]^{2-} + 2AsPh_3 \rightarrow (Ph_3As)_2Pd(SCN)_2$$
$$\rightarrow (Ph_3As)_2Pd(NCS)_2$$
$$[Pd(SCN)_4]^{2-} + bpy \rightarrow (bpy)Pd(SCN)_2 \rightarrow (bpy)Pd(NCS)_2$$

Similarly, K$_2$[Pd(SCN)$_4$] reacted with tetraethyldiethylentriamine (Et$_2$NCH$_2$CH$_2$NHCH$_2$CH$_2$NEt$_2$, Et$_4$dien) in aqueous solution at 0 °C:[15]

$$[Pd(SCN)_4]^{2-} + Et_4dien \rightarrow [(Et_4dien)PdSCN]^+$$

When the aqueous solution of [(Et$_4$dien)PdSCN]PF$_6$ was kept standing at room temperature, transformation to the N-bonded isomer occurred:

$$[(Et_4dien)PdSCN]^+ \rightarrow [(Et_4dien)PdNCS]^+$$

The reaction was followed by the change in the UV absorption spectrum exhibiting isosbestic points at 308 and 365 nm. The first-order rate constant ($k = 2.56 \times 10^{-3} \text{s}^{-1}$ at 45°C) and the kinetic parameters ($\Delta H^{\ddagger} = 17.3 \pm 0.5$ kcal mol^{-1} and $\Delta S^{\ddagger} = -16.8 \pm 1.0$ cal deg^{-1} mol^{-1}) coincide with those for the ligand substitution reaction:

$$[(Et_4 dien)PdSCN]^+ + Br^- \rightarrow [(Et_4 dien)PdBr]^+ + NCS^-$$

Hence, the thiocyanate linkage isomerization was thought to follow an intermolecular mechanism (dissociation and religation of NCS$^-$) in contrast with the nitrito-nitro case.[15] Later studies at high pressure (1–1500 bar) also suggest that both the linkage isomerization of and bromide substitution in $[(Et_4 dien)PdSCN]^+$ proceed via $[(Et_4 dien)Pd(H_2 O)]^{2+}$.[16]

On the other hand, linkage isomerization of $[(NH_3)_5 CoSCN]^{2+}$ in basic solution containing KS^{14}CN was not accompanied by the thiocyanate exchange.[17] Similarly, $[(H_2 O)_5 CrSCN]^{2+}$ isomerized to the N-bonded isomer in the presence of ^{35}S-labeled ammonium thiocyanate without appreciable ligand exchange.[18] These results suggest that an intramolecular mechanism is operative for the thiocyanate isomerization in these complexes. Alternatively, if the M—SCN bond is cleaved in the transition state, the intimate ion pair $L_5 M^{3+} \cdot SCN^-$ may undergo dissociation into $L_5 M^{3+}$ and SCN$^-$ relatively slowly compared with the rate of internal return.

Besides the nitrite,[6] cyanate, thiocyanate, and selenocyanate anions,[4] various other ligands, especially large organic ligands inclusive of chelating ligands, have been involved in linkage isomers.[5] Furthermore, in recent years novel types of isomers have been prepared in the field of organometallic chemistry. For example, the following isomers were prepared by the reactions of tetrakis(triphenylphosphine)palladium(0) with 2-, 3-, and 4-bromopyridines in toluene at 90°C followed by treatment of the products with triethylphosphine in diethyl ether at room temperature.[19]

Each of these compounds involves a σ-bonded pyridyl ligand which was deprotonated at a different ring carbon. They were characterized by ^{1}H and ^{13}C NMR spectroscopy and their molecular structures were confirmed by X-ray crystallography.[19]

2 Monoatomic Ligands

A great number of metal complexes containing monoatomic ligands such as H^-, halides, chalcogenides, N^{3-}, P^{3-}, and As^{3-} are known at present. Among them metal hydride complexes have been investigated most extensively, since they are not only interesting from a structural point of view, but also important as homogeneous catalysts or intermediates in hydrogenation and other reactions of organic substrates.

2.1 Coordination Modes for the Hydride Ligand[20-28]

It is now becoming common to locate H atoms directly using X-ray diffraction methods, while neutron diffraction offers distinct advantages. Since H atoms scatter neutrons with about the same efficiency as do most elements, their positions are determined with high precision. Two distinct disadvantages of neutron diffraction are: the necessity of a source of neutrons and of large single crystals.

The first hydride complex investigated by the neutron diffraction technique was $K_2[ReH_9]$ containing the terminal M—H linkage. Many examples of complexes containing the terminal hydride ligands are now known for virtually all d-block transition elements. Binary transition-metal hydrides are rather few and the majority are stabilized by carbonyl, phosphine, or other ancillary ligands.

Numerous structures containing the M—H—X linkage have also been elucidated for X = B, C, Si, Al, and other metal atoms. Dinuclear complexes containing the $M(\mu\text{-}H)_nM$ linkage ($n = 1-4$) aroused the structural interest. Furthermore, μ_2(edge-bridging), μ_3(face-bridging), and encapsulated (interstitial) hydride ligands have been found in metal clusters*.[22]

---

* Polynuclear compounds containing at least three metal atoms linked directly by covalent bonds are called *metal clusters*.[29-31]

2.1.1 Terminal Hydride Ligands

2.1.1.1 Preparative Methods

Generally, terminal-hydride complexes are prepared from metal halides, their phosphine derivatives, or organometallic complexes either by the action of a hydridometallate or dihydrogen in THF, alcohols, or other solvents. Potassium, sodium, sodium amalgam, and magnesium are also used as reducing agents. Some examples of preparative reactions are shown below:[26, 27]

$$mer\text{-}IrCl_3(AsEt_2Ph)_3 + LiAlH_4 \rightarrow fac\text{-}IrH_3(AsEt_2Ph)_3$$
$$CoCl_2 \cdot 6H_2O + PEtPh_2 + NaBH_4 \rightarrow CoH_3(PEtPh_2)_3$$
$$cis\text{-}PtCl_2(PEt_3)_2 + H_2 \rightarrow trans\text{-}PtHCl(PEt_3)_2$$
$$WMe_6 + PR_3 + H_2 \rightarrow WH_6(PR_3)_3$$
$$MoCl_4(PEt_2Ph)_2 + PEt_2Ph + Na + H_2 \rightarrow MoH_4(PEt_2Ph)_4$$
$$NaReO_4 + Na + EtOH \rightarrow Na_2ReH_9$$

Hydrolysis of complex salts and oxidative addition (p. 27) of weak acids are also utilized frequently:

$$Na_2[Fe(CO)_4] + H_2SO_4 \rightarrow FeH_2(CO)_4$$
$$Na[Rh(PF_3)_4] + H_2SO_4 \rightarrow RhH(PF_3)_4$$
$$RhCl(PPh_3)_3 + HSiCl_3 \rightarrow RhHCl(SiCl_3)(PPh_3)_2$$
$$Pt(PPh_3)_3 + HCN \rightarrow PtH(CN)(PPh_3)_2$$

2.1.1.2 Characterization

Spectroscopic techniques, especially infrared and nuclear magnetic resonance studies are very important in transition-metal hydride chemistry. The IR spectra of terminal-hydride complexes show the M—H stretching and bending modes in the 2250–1700 and 800–600 cm^{-1} regions, respectively.[32] The v(M—H) band is relatively sharp and of medium intensity, affording a diagnostic information for hydride ligands. It is particularly valuable for complexes unsuited to NMR spectroscopy because of paramagnetism or low solubility. The v(M—H) frequency reflects the nature of the ligand in the *trans* position in square-planar complexes: ligands with high *trans* influence weaken the M—H bond, producing a concomitant reduction in the v(M—H) frequency. Thus, the order of v(Pt—H) in *trans*-PtHX(PEt$_3$)$_2$ is as follows (in order of increasing *trans* influence of X):[32]

X =	NO$_3^-$ < Cl$^-$ < Br$^-$ < I$^-$ < NO$_2^-$ < SCN$^-$ < CN$^-$
v(Pt—H)/cm^{-1}	2242 > 2183 > 2178 > 2156 > 2150 > 2112 > 2041

More reliable evidence for the presence of hydride ligands is provided by ^{1}H NMR spectroscopy. For terminal hydride ligands, resonances are observed typically in the high field between -5 to -25 ppm away from tetramethylsilane.[26] This is attributed to the strong shielding of the ^{1}H nuclei by the d electrons of the central metal. Furthermore, the relative intensities of ^{1}H signals are utilized in determining the stoichiometry of polyhydrides.

When hydride complexes contain phosphorus donors, the ^{1}H signal is spin-spin coupled to ^{31}P, which reveals some information concerning the geometrical arrangements. The ^{31}P NMR spectra are also helpful.[33] Similarly, the values of coupling constant $J(^{195}\text{Pt}-\text{H})$ in platinum hydride complexes afford useful data. Thus, $J(\text{Pt}-\text{H})$ and $\nu(\text{Pt}-\text{H})$ in the $trans$-PtHL(PEt$_3$)$_2$ complexes decrease in the same order:

L =	py	<	CO	<	PPh$_3$	<	P(OPh)$_3$	<	P(OMe)$_3$	<	PEt$_3$
$J(\text{Pt-H})/\text{Hz}$	1106	>	967	>	890	>	872	>	846	>	790
$\nu(\text{Pt-H})/\text{cm}^{-1}$	2216	>	2167	>	2100	>	2090	>	2067	<	2090

where the σ-donor strength of L increases as the $J(\text{Pt}-\text{H})$ value decreases and $\nu(\text{Pt}-\text{H})$ shifts to a lower frequency.[32]

X-Ray and neutron diffraction analyses are the most reliable techniques, giving both accurate hydride-ligand positions and M—H distances. Some examples of terminal-hydride complexes whose structures were determined by these methods are described below with reference to their molecular geometries.

2.1.1.3 Tricapped Trigonal Prism Structure

Potassium nonahydridorhenate(VII) K$_2$[ReH$_9$] was prepared by reduction of potassium perrhenate in aqueous ethylenediamine with potassium or lithium metal. Neutron diffraction analysis established unequivocally the correct composition and molecular structure of this polyhydride complex. The diffraction study was performed within an evacuated cryostat maintained at room temperature, since the crystals are unstable in moist air.[34] The Re atom lies at the center of a trigonal prism of H atoms, with three H atoms beyond the centers of the prism faces. Thus, the structure is a tricapped trigonal prism of D_{3h} symmetry. The average Re—H distance is 1.68(1) Å and the average H—Re—H angle between hydrogens in the same vertical mirror plane is 93.6(6)°.

Another example of the nine-coordinate polyhydride is WH$_6$ {PPh(i-Pr)$_2$}$_3$ which was prepared by subjecting a hexane solution of hexamethyltungsten and PPh(i-Pr)$_2$ to hydrogen at 300 atm for five days. The molecular structure was also confirmed by neutron diffraction to be a tricapped trigonal prism in which the phosphine ligands occupy a pair of eclipsed prism corners and one of the equatorial sites (Fig. 1).[35] The molecule has approximate C_{2v} symmetry with W—H distances in the range of 1.718(3)–1,745(3) Å; the W—P(2) distance of 2.424(2) Å is remarkably shorter than the other two W—P separations (average 2.522(2) Å).

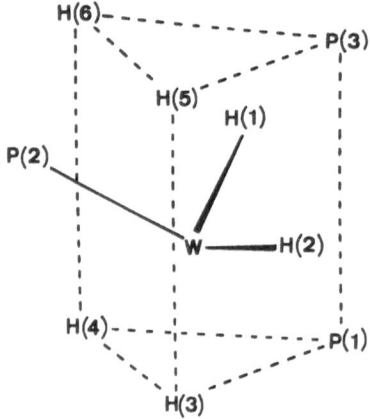

Fig. 1. Coordination sphere of the molecule $WH_6\{PPh(i\text{-}Pr)_2\}_3$ whose structure was determined by neutron diffraction at 100 K.[35]

2.1.1.4 Octahedral Structure

A binary metal hydride anion $[FeH_6]^{4-}$, which is isoelectronic with $[ReH_9]^{2-}$, was found by X-ray analysis to exist in a crystal of $Fe_6H_6Mg_4Br_{3.5}Cl_{0.5}(THF)_8$.[36] The H—Fe—H angles in the range of 88(4)–92(4)° conform well with octahedral coordination symmetry with the average Fe—H distance of 1.69(9) Å. This compound was obtained from reaction of $FeCl_3$ with phenylmagnesium bromide in the presence of excess hydrogen followed by extraction of the product with THF, and is extremely sensitive to air and moisture.

2.1.1.5 Pentagonal-Bipyramid Structure

Neutron diffraction studies of $OsH_4(PMe_2Ph)_3$ and $IrH_5\{P(i\text{-}Pr)_3\}_2$ at 90 and 80 K, respectively, established the pentagonal-bipyramid geometries around Os and Ir. In the former complex, Os, one P, and four H, atoms lie on the equatorial plane with the Os—H length ranging from 1.644(3) to 1.681(3) Å and the H—Os—H angle being 67.9(2)–70.0(2)°. The *trans* P—Os—P and average *cis* P—Os—P angles are 166.1(1) and 97.0(1)°, respectively.[37] The latter Ir(V) complex was recently prepared by reduction of $IrHCl_2\{P(i\text{-}Pr)_3\}_2$ with $LiAlH_4$ in THF. It has five equatorial H atoms, the average Ir—H distance and the H—Ir—H angle being 1.603(9) Å and 71.9(11)°, respectively. The P—Ir—P axis is linear, and the six isopropyl-carbon atoms are staggered with respect to this axis.[38]

2.1.1.6 *Trans* Influence of the Hydride Ligand

In spite of the variety in the coordination geometries, the length of a terminal M—H bond is in the range of 1.60–1.75 Å, consistent with that expected for a normal covalent linkage. The hydride ligand is well known to exert a strong *trans* influence in square-planar and octahedral complexes.[39] Thus, the structure of $K_3[RhH(CN)_5] \cdot H_2O$ determined by X-ray analysis (Fig. 2) shows the higher *trans* influence of H as compared to the cyanide ligand, the Rh-C distance *trans* to H being 0.08 Å longer than the four equatorial ones.[40]

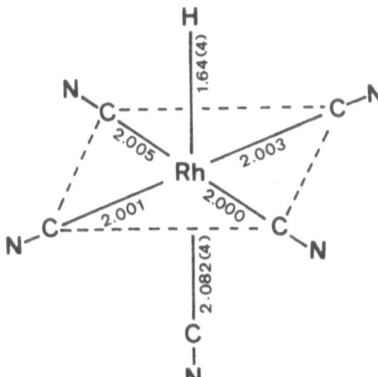

Fig. 2. X-Ray structure of the $[RhH(CN)_5]^{3-}$ ion.[40]

Therefore two hydride ligands usually avoid to occupy mutually *trans* coordination sites. For example, three H ligands in [K(18-crown-6)] $[RuH_3(PPh_3)_3]$ lie at the facial positions of a highly distorted octahedron.[41] It seems unusual that $Ir(H)_2(Cl)_2\{P(i\text{-}Pr)_3\}_2$ has three pairs of *trans* ligands. The Ir—H distance is quite large (Fig. 3).[42]

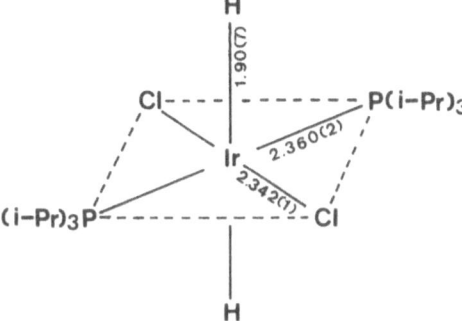

Fig. 3. Molecular structure of $Ir(H)_2Cl_2\{P(i\text{-}Pr)_3\}_2$.[42]

Paramagnetic transition-metal hydride complexes are very rare, and this 17-electron Ir(IV) complex is the first example of a stable paramagnetic hydride complex of the platinum-group metals. It was prepared by the reaction of $(NH_4)_2[IrCl_6]$ with an excess amount of $P(i\text{-}Pr)_3$ in refluxing ethanol containing concentrated HCl.[42]

2.1.2 Unsupported M-H-M Linkage

Dinuclear complexes in which the two halves of the molecule are held together solely by one three-center two-electron (3c–2e) bond without other bridging groups are quite rare. The $[HM_2(CO)_{10}]^-$ family (M = Cr, Mo, or W) and close analogs such as $HW_2(CO)_9(NO)$ are among the very few transition-metal complexes known to contain such an "unsupported" M—H—M bond.

There are four possible structures: linear/eclipsed; linear/staggered; bent/eclipsed; and bent/staggered. They are defined according to the linearity (or lack thereof) of the intersecting vectors (axial)OC—M $\cdots$ M—CO(axial) and relative orientations (eclipsing or staggering) of the two equatorial $M(CO)_4$ or $M(CO)_3L$ units. Preference for a structure is made by the nature of M, of the counterion, and of substituent ligands. The bent/staggered form predominates in the known solid-state structures. The second most common is the linear/eclipsed form and there are no examples reported thus far of the bent/eclipsed configuration. As is seen in Fig. 4, the anion $[HCr_2(CO)_{10}]^-$ in crystals of the $[Et_4N]^+$ salt has a linear/eclipsed structure of D_{4h} symmetry. The H atom is "off-axis"

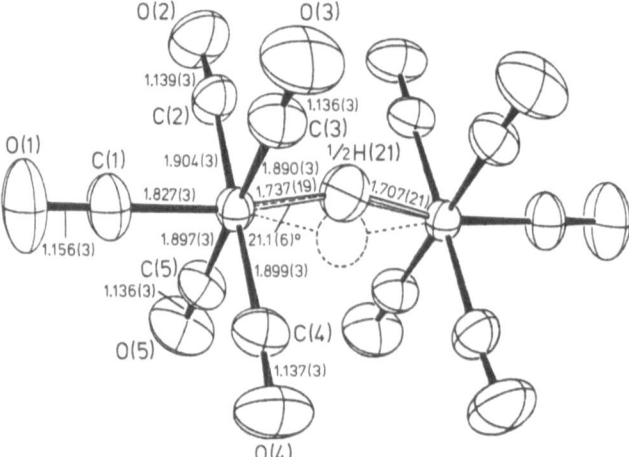

Fig. 4. Structure of the anion in $[Et_4N][HCr_2(CO)_{10}]$ based on the neutron diffraction data.[43]

(0.3 Å from the center of the Cr—Cr bond), resulting in a bent Cr—H—Cr linkage (158.9(6)°).[43] The anions in the bis(triphenylphosphine)nitrogen(1+) salts: $[(Ph_3P)_2N][HCr_2(CO)_{10}]$[44] and $[(Ph_3P)_2N]$ $[DCr_2(CO)_{10}]$[45] also have the linear/eclipsed configuration.

In contrast, the structure of the $[(\mu\text{-}H)W_2(CO)_{10}]^-$ anion is affected appreciably by the nature of counterions and the influence of crystal packing. The nonhydrogen skeletons of the anions in $[Et_4N]$ $[HW_2(CO)_{10}]$ were found by X-ray analysis to be linear/eclipsed, whereas the anions in the $[(Ph_3P)_2N]^+$ salt are appreciably bent/staggered. The W—W separation in the bent form (3.391(1) Å) is 0.11 Å smaller than that in the linear form (3.504(1) Å).[46]

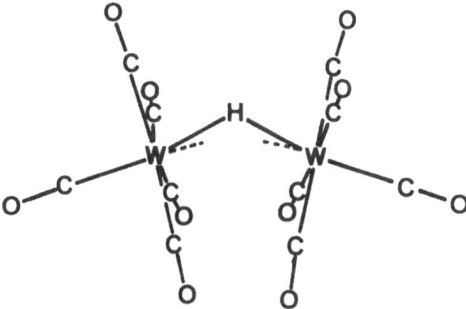

Fig. 5. A molecular plot of the anion in $[Ph_4P][HW_2(CO)_{10}]$ based on neutron diffraction data collected at 40 K.[47] The 45° staggering of the equatorial CO groups is observed.

Figure 5 shows a similar bent/staggered structure of the $[HW_2(CO)_{10}]^-$ anion in $[Ph_4P][HW_2(CO)_{10}]$ established by neutron diffraction analysis at 40 K.[47] The W—H and W—W distances are 1.897(5) and 3.340(5) Å, respectively. A crystallographic two-fold axis passes through the bridging H atom, bisecting the W—H—W angle (123.4(5)°). The axial O—C—W vector is not colinear with the bridging H atom, but is directed approximately toward the center of the WHW triangle.

Of the three 6B group $[(\mu\text{-}H)M_2(CO)_{10}]^-$ complexes, the molybdenum derivative is least stable both toward dimer dissociation and toward CO lability. X-Ray diffraction studies have revealed that the $[HMo_2(CO)_{10}]^-$ anion can adopt either a linear/eclipsed ($[Et_4N]^+$ salt) or an appreciably bent/staggered ($[(Ph_3P)_2N]^+$ salt) configuration.[48]

Related anions in

[Et$_4$N][HMo$_2$(CO)$_9$(PPh$_3$)],[49]
[Et$_4$N][HMo$_2$(CO)$_8$(PPh$_2$Me)$_2$],[50]
[Ph$_4$P][HMo$_2$(Co)$_8$(Ph$_2$PCH$_2$PPh$_2$)],[51]
and
[Et$_4$N][HMo$_2$(CO)$_8${Ph$_2$P(CH$_2$)$_4$PPh$_2$}][51]

also have the bent/staggered geometry.

   In all of the 6B group MHM carbonylates and derivatives, the H atoms are positioned off the M $\cdots$ M axis in linear arrangement or off the intersection point of the two OC(axial)—M vectors. These arrangements are consistent with the presence of closed three-center, two-electron bonds (Fig. 6) analogous to those found in borane chemistry and thus imply a certain degreee of metal-metal bonding.

Fig. 6 a, b. Orbital representation (**a**) and symbol (**b**) for the M—H—M three-center, two-electron bond.

2.1.3 Dinuclear M(μ-H)$_n$M and Mixed-bridged Systems

2.1.3.1 Preparative Methods

Mononuclear hydride complexes may act as donor ligands to donate one or more hydride atoms to a coordinatively unsaturated acceptor complex, affording dinuclear hydrogen-bridged species. Examples of these reactions, some of which involve the subsequent loss of dihydrogen from the adduct, are given below.[26] Here, S denotes a solvent molecule which is included for "book-keeping" of the coordination number:

[Pt(Ph)(PEt$_3$)$_2$S]$^+$ + H$_2$W(η-C$_5$H$_5$)$_2$
$\quad\quad\quad\quad \rightarrow$ [(PEt$_3$)$_2$(Ph)Pt(μ-H)$_2$W(η-C$_5$H$_5$)$_2$]$^+$
[Rh(H)$_2$(PPh$_3$)$_2$(S)$_2$]$^+$ + H$_2$W(η-C$_5$H$_5$)$_2$
$\quad\quad\quad\quad \rightarrow$ [(PPh$_3$)$_2$Rh(μ-H)$_2$W(η-C$_5$H$_5$)$_2$]$^+$ + H$_2$
[Rh(PPh$_3$)$_2$(S)$_2$]$^+$ + H$_3$Ir(PEt$_3$)$_3$
$\quad\quad\quad\quad \rightarrow$ [(PPh$_3$)$_2$Rh(μ-H)$_3$Ir(PEt$_3$)$_3$]$^+$

2.1.3.2 Characterization

Bridging hydride ligands are usually more firmly bound than their terminal counterparts, and are readily characterized by spectroscopic methods such as NMR and mass spectroscopy. Application of the NMR spectroscopy is sometimes difficult because of low solubility and inadequate relaxation times. When obtained, signals from the μ_2-bridging hydrides usually appear at higher field than those from the terminal hydrides.

Carbonyl hydride clusters are moderately volatile and subjected to mass spectroscopy. Whereas terminal hydride ligands are frequently lost upon ionization, bridging hydride ligands are tenaciously retained. Therefore, mass spectroscopy is useful in distinguishing between terminal and bridging hydride ligands.

2.1.3.3 M(μ-H)$_2$M Systems

Several dinuclear complexes containing the (μ-H)$_2$ bridges have been studied by neutron diffraction. Three representative structures are exemplified in Figs. 7, 8, and 9. The Rh complex is square planar,[52] while the coordination about each W atom in $[H_2W_2(CO)_8]^{2-}$ is approximately octahedral.[53] In either complex, both H atoms are situated somewhat inside the intersection of the *trans* P—Rh or OC—W vectors. This may be contrasted with the M(μ-H)M cases where H atoms are displaced outside the intersection of two OC(axial)—M vectors. Fur-

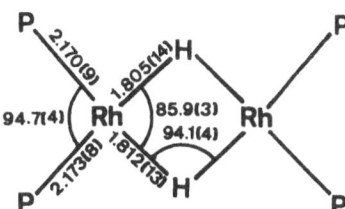

Fig. 7. The inner core of $[HRh\{P(O\text{-}i\text{-}Pr)_3\}_2]_2$ with selected bond distances (Å) and angles which were determined by a neutron diffraction study at 112.4 K. Rh—Rh = 2.647(13) Å and H $\cdots$ H = 2.465(14) Å.[52]

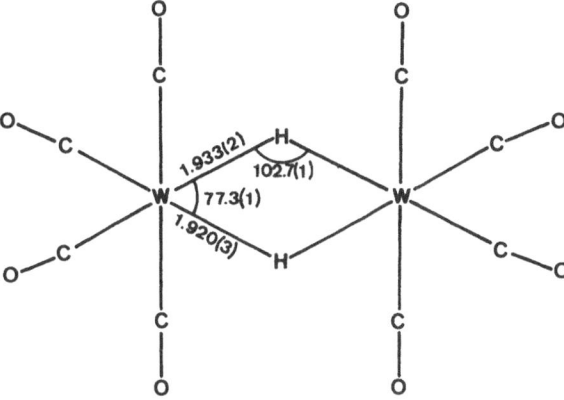

Fig. 8. Structure of the anion in $[(Ph_3P)_2N]_2[H_2W_2(CO)_8]$ determined by neutron diffraction technique at 28 K. W—W = 3.010(2) Å and H $\cdots$ H = 2.405 Å.[53]

thermore, the average W—H length in $[H_2W_2(CO)_8]^{2-}$ (1.926(2) Å) is significantly longer than that (1.897(5) Å) in $[Ph_4P][HW_2(CO)_{10}]$, indicating that the M—H bond order is lower in the $M(\mu\text{-}H)_2M$ system than in the $M(\mu\text{-}H)M$ linkage.

As is seen in Fig. 9, the cation in $[H_3Pt_2(dppe)_2][BPh_4]$ (dppe = $Ph_2PCH_2CH_2PPh_2$) has a unique structure.[54] The five-coordinate Pt(1) atom has a distorted trigonal-bipyramid geometry, with H(t) and P as the axial atoms, with the Pt—H(t) length being 1.610(2) Å. The four-coordinate Pt(2), on the other hand, has a normal square-planar geometry. The most noticeable feature of the two bridging H atoms is the marked assymmetry of their positions. They interact much more strongly with Pt(2) than with Pt(1), and are more "terminal" in nature than "bridging".

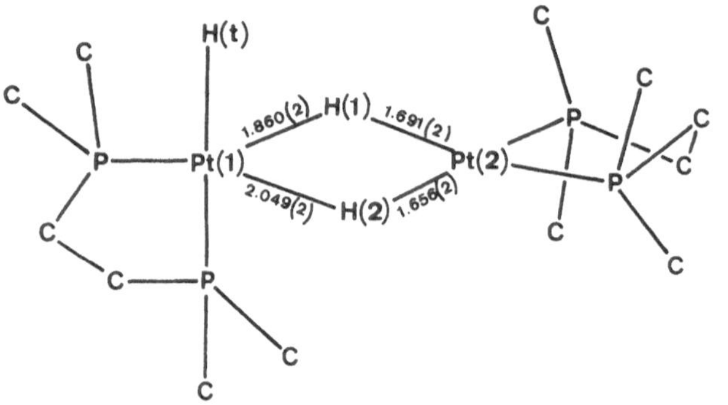

Fig. 9. Structure of the cation in $[H_3Pt_2(dppe)_2][BPh_4]$ determined by neutron diffraction technique at 20 K. Bond angles (°): Pt(1)H(1)Pt(2) 99.39(11), Pt(1)H(2)Pt(2) 93.41(10), H(1)Pt(1)H(2) 75.53(9), H(1)Pt(2)H(2) 91.55(11); Pt $\cdots$ Pt = 2.711(1) Å, H(1) $\cdots$ H(2) = 2.399(3) Å.[54]

2.1.3.4 $M(\mu\text{-}H)_3M$ Systems

Dinuclear $M(\mu\text{-}H)_3M$ compounds are rare. One example is the terminal-hydride complex $RhH_3(triphos)$ (triphos = $MeC(CH_2PPh_2)_3$) which reacts with an equimolar amount of HSO_3CF_3 in THF evolving H_2 and producing an orange solution from which crystals of $[(triphos)HRh(\mu\text{-}H)_2RhH(triphos)][BPh_4]_2$ have been obtained. An N,N-dimethyl-formamide (DMF) solution of this complex was exposed to air for 12 h and a n-BuOH solution of $NaBPh_4$ was added to yield red crystals of $[(triphos)Rh(\mu\text{-}H)_3Rh(triphos)][BPh_4]_2 \cdot DMF$.[55]

The structure of this paramagnetic Rh(III)–Rh(II) complex was determined by the X-ray method. Two octahedra around the Rh metals share a trigonal face $(\mu\text{-H})_3$, and the two (triphos)Rh units are rotated ca. 15° from a mutually eclipsed orientation. Average bond lengths and angles are Rh(1)—H 1.73(6) Å, Rh(2)—H 1.91(7) Å, Rh(1)–Rh(2) 2.644(1) Å, H—Rh(1)—H 77(3)°, and H—Rh(2)—H 67(3)°. The magnetic moment of this compound (2.20 Bohr magneton) corresponds to one unpaired spin, and the Rh—Rh distance suggests the existence of a multiple bond although the exact bond order is not certain.[55]

Similar triply hydrogen-bridged structures have also been disclosed by X-ray analysis for $[(t\text{-BuNC})(PPh_3)_2 HRe(\mu\text{-H})_3 ReH(PPh_3)_2 (t\text{-BuNC})]PF_6$[56] and $[Et_4N][(triphos)Re(\mu\text{-H})_3 ReH_4] \cdot MeCN$.[57]

The structure of a cation in the former compound may be described as two face-sharing distorted pentagonal bipyramids, two $ReH(PPh_3)_2$ $(t\text{-BuNC})$ halves joining by three bridging hydrogen atoms, and a Re—Re bond (2.604(1) Å). The Re—H (bridging) length ranging from 1.65(9) to 1.99(7) Å is appreciably longer than the Re—H (terminal) (1.50(8) and 1.68(7) Å).[56]

On the other hand, the dinuclear anion in the latter compound has an unsymmetrical structure. One Re atom is coordinated with four terminal and three bridging hydrogen atoms, defining a distorted pentagonal bipyramid. The average Re—H(terminal) and Re—H(bridging) distances are 1.59(14) and 1.77(20) Å, respectively. The other Re atom caps the trigonal plane which is defined by the bridging hydrogen atoms, with the average Re—H(bridging) length being 1.85(7) Å. The three phosphorus atoms of the triphos molecule complete a slightly distorted octahedron around the second Re atom. The two Re atoms are separated by 2.597(1) Å and are considered to be linked by a triple bond.[57]

2.1.3.5 $M(\mu\text{-H})_4 M$ Systems

A quadruply hydrogen-bridged metal-metal bond was established by the neutron diffraction method at 80 K for $Re_2H_8(PEt_2Ph)_4$ which was prepared by reduction of $[Re_2Cl_8]^{2-}$ with $NaBH_4$ in EtOH solution containing PEt_2Ph. As is seen in Fig. 10, a crystallographic center of

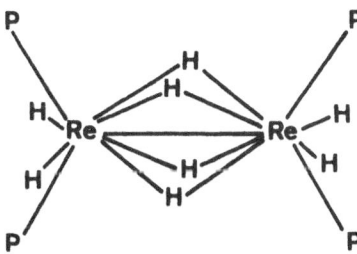

Fig. 10. The molecular geometry of $H_8Re_2(PEt_2Ph)_4$ which was determined by neutron diffraction at 80 K.[58]

symmetry exists at the midpoint of the Re—Re bond (2.538(4) Å). Four bridging H atoms are situated at an average distance of 1.38 Å from this midpoint, defining a distorted square plane which is essentially perpendicular to the Re—Re bond. Thus, the H_4Re_2 core is almost octahedral. The terminal H_2P_2 units and the bridging H_4 group are in a mutually staggered arrangement. Average Re—H (bridging) und Re—H (terminal) distances are 1.878(7) and 1.669(7) Å, respectively, and the Re—Re bond length is 2.538(4) Å.[58]

2.1.3.6 M(μ-H)(μ-X)$_n$M Systems

Rather many homo- and hetero-dinuclear metal complexes containing mixed (μ-H)(μ-X) bridges have been studied. Among recent examples are

$(\eta^4\text{-}C_8H_{12})Ir(\mu\text{-}H)(\mu\text{-}Cl)IrH_2(PPh_3)_2$,[59]
$[pyH]_3[Cl_3Mo(\mu\text{-}H)(\mu\text{-}Cl)_2MoCl_3]$,[60]
$[Ph_4P][Cl_3W(\mu\text{-}H)(\mu\text{-}Me_2S)_2WCl_3]$,[61]
$[(dppe)Rh(\mu\text{-}H)(\mu\text{-}Cl)IrCl(PEt_3)_3][BF_4]$,[62]

and

$(\eta^4\text{-}C_8H_{12})Rh(\mu\text{-}H)(\mu\text{-}PhPCH_2PPh_2)Ru(Ph)(dppm)$
$(dppm = Ph_2PCH_2PPh_2)$.[63]

2.1.4 Edge-bridging (μ_2) and Face-bridging (μ_3) Hydride Ligands in Metal Clusters

Many metal clusters containing edge-bridging (μ_2) hydride ligands have been prepared from cluster precursors by a variety of specific reactions. These include protonation, oxidative addition (p. 27) of dihydrogen, activated hydrocarbons or functional organic molecules, cyclometallations, reactions with water and other hydrogen compounds, and reduction by base or complex metal hydrides.[22] Some examples are given below:

$$Ru_3(CO)_{12} + \text{conc. } H_2SO_4 \rightarrow [Ru_3H(CO)_{12}]^+ + HSO_3^-$$
$$Os_3(CO)_{12} + H_2 \rightarrow Os_3H_2(CO)_{10} + 2CO$$
$$Os_3(CO)_{12} + C_2H_4 \rightarrow Os_3H_2(C{=}CH_2)(CO)_9 + 3CO$$
$$Ru_3(CO)_{12} + H_2O \rightarrow Ru_4H_2(CO)_{13}, Ru_4H_4(CO)_{12}$$

As an example the molecular skeleton of $Ru_4(\mu\text{-}H)_4(CO)_8$ $\{P(OMe)_3\}_4$ is drawn schematically in Fig. 11. Each Ru atom is

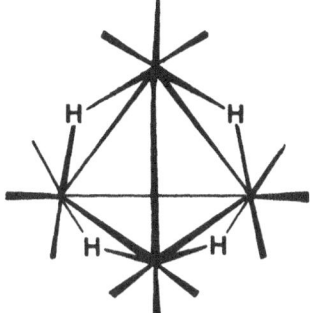

Fig. 11. Schematic drawing of the molecule of $H_4Ru_4(CO)_8\{P(OMe)_3\}_4$ whose structure was determined by neutron diffraction at 20 K.[64]

approximately octahedrally coordinated by one $P(OMe)_3$, two CO, one Ru atom, and two H ligands which bridge to one further Ru atom each.

According to the neutron diffraction results,[64] all the Ru—H—Ru bridges are symmetrical with a mean Ru—H length of 1.773(2) Å and mean a bond angle of $<$Ru—H—Ru $= 114.2(3)°$. The Ru—H—Ru bonds are regarded as closed 3c-2e bonds (Fig. 6). The Ru_4 unit constitutes a distorted tetrahedron with four long, H-bridged vectors with mean lengths of 2.978(4) Å and two shorter, unbridged Ru—Ru bonds with a mean length of 2.792(6) Å.

The face-bridging (μ_3) mode of hydrogen binding to a metal cluster is much less common than the edge-bridging mode. The cation in $[H_4Rh_4(\eta\text{-}C_5Me_5)_4][BF_4]_2$ presents a recent example, whose structure was established by neutron diffraction at 12 K. As is seen in Fig. 12, the

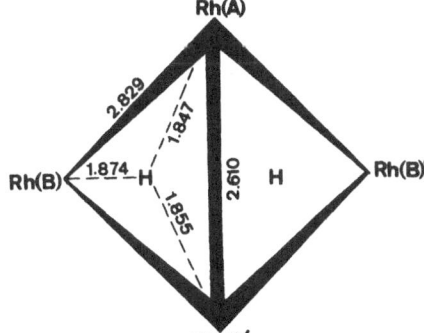

Fig. 12. The tetranuclear metal core in $[H_4Rh_4(\eta\text{-}C_5Me_5)_4][BF_4]_2$ as revealed by neutron diffraction analysis at 12 K. Esd's are 0.005 or 0.006 Å.[65]

Rh_4 core shows substantial distortion from a regular tetrahedron, resulting in four long and two short Rh—Rh distances. The four hydride ligands bridge the faces of the Rh_4 cluster at a distance of 0.957 Å above each plane formed by the three Rh atoms. Each C_5 plane is 1.838 Å from the rhodium to which it is bonded.[65]

2.1.5 Systems with an Interstitial Hydrogen Atom

Several metal clusters have been assumed to contain an interstitial H atom, but its direct location by neutron diffraction analysis was achieved only for two complexes, $[(Ph_3P)_2N][HCo_6(CO)_{15}]$ and $[Ph_4As]$ $[HRu_6(CO)_{18}]$. In either compound a six-coordinate H atom is situated in the geometric center of the octahedral cluster.

The cobalt complex was derived from the precursor cluster $K_2[Co_6(CO)_{15}]$ by reaction with concentrated hydrochloric acid, followed by metathesis of the product with $[(Ph_3P)_2N]Cl$ in 2-propanol. As is seen in Fig. 13, the Co complex contains ten terminal and five bridging CO groups. The mean length of the bridged Co—Co bonds is 2.532(9) Å and that of the non-bridged Co—Co bonds is 2.613(15) Å. Both distances are larger than the average Co—Co distance (2.51 Å) in the precursor anion $[Co_6(CO)_{15}]^{2-}$, showing significant expansion of the octahedral core upon insertion of a proton. The average Co—H distance is 1.823(13) Å and the *trans* Co—H—Co bonds are very nearly linear (mean angle 176.2(9)°).[66]

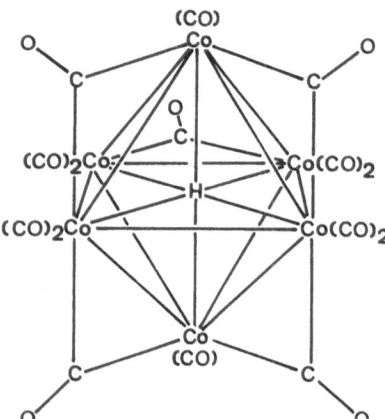

Fig. 13. Structure of the $[HCo_6(CO)_{15}]^-$ anion in the $[(Ph_3P)_2N]^+$ salt as determined by neutron diffraction analysis at 80 K.[66]

The complex anion in $[Ph_4As][HRu_6(CO)_{18}]$ contains no external bridge, all carbonyl ligands being terminal. The hydrogen atom lies at the center of the octahedron and the Ru—H distance ranges from 2.028(11) Å to 2.055(12) Å averaging 2.037(12) Å.[67]

The interstitial H atoms in these clusters exhibit anomalously low NMR chemical shifts. Thus, the proton signal from $[HCo_6(CO)_{15}]^-$ appears at δ 23.2 ppm in acetone-h_6 at 173 K; it disappears, however, upon warming to room temperature. Some type of exchange of the proton seems to occur in solution, since the cluster anion shows no proton signal in acetone-d_6 and is readily deprotonated by water, methanol, and other proton-accepting solvents.[66]

On the other hand, $[HRu_6(CO)_{18}]^-$ shows a proton singlet at δ 16.43 ppm (the $[(Ph_3P)_2N]^+$ salt in CD_2Cl_2 at 40°C) or at δ 16.49 ppm (the $[Me_4N]^+$ salt in THF-d_8 at 40°C). However, the behavior of the interstitial H atom in the Ru cluster is different from that in the Co cluster. Neither excess KOH in methanol nor excess KH in THF could deprotonate the Ru cluster, a large excess of base causing decomposition with loss of the carbonyl IR spectrum.[68]

2.1.6 Metal Clusters Including an Interstitial Light Atom Other than Hydrogen

In recent years a great number of transition metal clusters containing an interstitial carbon atom has been prepared and now they constitute a growing field of "metal carbide clusters".[69, 70] A far less number of nitride clusters[71] and several compounds containing a naked P, As, or S atom embedded within metal clusters have been reported, such as

$[Et_3(PhCH_2)N]_3[Ru_5N(CO)_{14}]^{72}$
$[Et_3(PhCH_2)N]_3[Rh_{10}P(CO)_{22}],^{73}$
$[Et_3(PhCH_2)N]_3[Rh_{10}As(CO)_{22}],^{74}$
and
$[(Ph_3P)_2N]_2[Rh_{10}S(CO)_{22}].^{75}$

There are, however, profound differences between the chemical behaviors of interstitial H and other atoms. In particular, no reaction analogous to the easy exchange of protons between the $[HCo_6(CO)_{15}]^-$ anions and their surroundings is found in other interstitial complexes. Many more interstitial hydride clusters would be desirable.

2.2 Chemical Reactions of Hydride Ligands

As to the hydride-protonic preference of hydride compounds, those of the representative metals are hydridic, while those of the non-metals are protonic. The transition metals are in a boarderline situation. In general, hydrides which are good proton donors, are poor hydride donors. The best hydride donors are found to the left in the periodic table. Thus, the hydrides of Ti, Zr, and Hf show no acid properties, but behave as typical hydride ion donors.[28]

Table I. Brønsted Acidity of Transition-Metal Hydrides[28]

Hydride	pK_a
Water, 25°C	
$HCo(CO)_4$	strong acid
$HCo(CO)_3(PPh_3)$	7.0
$HCo(CO)_3\{P(OPh)_3\}$	5.0
$HMn(CO)_5$	7.1
$H_2Fe(CO)_4$	$4.4 = pK_1$
	ca. $14 = pK_2$
$HFe(NO)(CO)_3$	ca. 5.1
$HV(CO)_6$	strong
$HV(CO)_5(PPh_3)$	6.8
$HM(PF_3)_4$, (Co, Rh, Ir)	strong
$HRe(CO)_5$	very weak
$HCo(dmgH)_2(PBu_3)$	10.5
$HRh(dmgH)_2(PPh_3)$	9.5
$[HCo(CN)_5]^-$	ca. 20
Methanol, 25°C	
$H_4Ru_4(CO)_{12}$	11.9
$H_4FeRu_3(CO)_{12}$	13.4
$H_2Ru_4(CO)_{13}$	14.7
$H_2FeRu_3(CO)_{12}$	14.3
$H_2Os(CO)_4$	14.7
$H_4Os_4(CO)_{12}$	12.0
$H_2Os_3(CO)_{12}$	14.5
$H_2Fe(CO)_4$	6.8
$H_4Ru_4(CO)_{11}P(OMe)_3$	13.6
$HCr(\eta\text{-}C_5H_5)(CO)_3$	6.4
$HMo(\eta\text{-}C_5H_5)(CO)_3$	7.2
$HW(\eta\text{-}C_5H_5)(CO)_3$	9.0
$[HNi(dppe)_2]^+$	2.6
$[HNi\{P(OMe)_3\}_4]^+$	1.5
$[HPd\{P(OMe)_3\}_4]^+$	0.7
$[HPt\{P(OMe)_3\}_4]^+$	10.2
$[HIr(CO)Cl(PPh_3)_2]^+$	2.1
$[HRh(CO)Cl(PPh_3)_2]^+$	1.8
$[HRh(dppe)(MeOH)_2]^{2+}$	1.0
$[HIr(CO)Br(PPh_3)_2]^+$	2.6
$[HIr(CO)I(PPh_3)_2]^+$	2.8

Table I lists the Brønsted acidity of transition metal hydrides compiled by Pearson.[28] Acidity increases in going from left to right in a given transition series: $HMn(CO)_5 < H_2Fe(CO)_4 < HCo(CO)_4$. In general, the third transition series gives the least acidic hydrides. The second transition series is less acid than the first for the metals to the left in the periodic table, but the pattern is changed to the right. Thus the order of decreasing acidity is $Cr > Mo > W$ for $HM(\eta\text{-}C_5H_5)(CO)_3$, but $Rh > Co > Ir$ for $HM(CO)_4$.[28]

Reactions of individual hydride complexes vary widely and some representative types of reactions of terminal hydrides are described here only as examples.[26] The most important reaction of M—H bonds is that with unsaturated substances containing C=C or C≡C bonds. The reactions with olefins will be discussed in detail later.

2.2.1 Reactions with Acids

Hydride complexes react with a variety of acids. Strong acids usually result in salts, while weak acids form oxidative addition products. In the latter case, reductive elimination of dihydrogen may follow, affording a convenient route for introducing conjugate bases X such as C≡CR, SiR_3, acac (aceylacetonate), SPh, and OCOR into the coordination sphere of the metal:

$$IrH(CO)(PPh_3)_3 + HClO_4 \rightarrow [IrH_2(CO)(PPh_3)_3]ClO_4$$
$$IrH(CO)(PPh_3)_3 + HSnMe_3 \rightarrow Ir(H)_2(SnMe_3)(CO)(PPh_3)_2 + PPh_3$$
$$MHL_n + HX \rightarrow MH_2XL_n \rightarrow MXL_n + H_2$$

Triphenylmethyl tetrafluoroborate abstracts a hydride ligand:

$$RuH_2(PPh_3)_4 + (Ph_3C)BF_4 \rightarrow [RuH(PPh_3)_4]BF_4 + Ph_3CH$$

2.2.2 Reactions with Bases

Acidic hydrides react with strong bases such as alkyl or aryl lithium and alkali hydrides to afford anionic complexes:

$$ReH(\eta\text{-}C_5H_5)_2 + LiBu \rightarrow Li[Re(\eta\text{-}C_5H_5)_2] + BuH$$
$$CoH\{P(OMe)_3\}_4 + KH \rightarrow K[Co\{P(OMe)_3\}_4] + H_2$$

2.2.3 Reactions with Halogens and Organic Halides

Most transition-metal hydrides react readily with halogens. For example, an excess of I_2 totally decomposes $ReH_5(PPh_3)_3$ to ReO_2, but stoichiometric amounts of I_2, Br_2, or $SnCl_2$ give halido-tetrahydrido complexes:

$$2\,ReH_5(PPh_3)_3 + X_2 \rightarrow 2\,ReH_4X(PPh_3)_3 + H_2$$

Organic halides can also act as halide sources. Many polyhydrides decomopose in halogenated solvents over time to yield the corresponding metal halido complexes. The reactivity of the halogenated hydrocarbon increases with the halogen content:

$$CH_3Cl < CH_2Cl_2 < CHCl_3 < CCl_4.$$

2.2.4 Intramolecular Migratory Insertion

There is a substantial body of organometallic reactions which are called "insertion" reactions.[76, 77] This is a superficial misleading designation; the real process is composed of the intramolecular migration of a ligand onto an adjacent one followed by ligation of a solvent molecule or another ligand into the vacated coordination site:

$$\begin{array}{ccc} H & & L \\ | & & | \\ M\!-\!X + L & \rightarrow & M\!-\!X\!-\!H \end{array}$$

Hydrogen migration onto the adjacent olefin ligand is a key step in hydrogenation reactions and will be treated in the succeeding section.

2.2.5 Reductive Elimination Reactions

Elimination of H_2, RH, or HX from a metal hydride complex may occur when the metal atom can adopt a stable oxidation state two units below that in the parent complex. Reductive elimination of dihydrogen can readily be achieved by thermal or photochemical methods, and is promoted by the presence of a donor ligand capable of reacting with the product. For example, the following reactions proceed photochemically:

$$MoH_4(dppe)_2 + 2N_2 \rightarrow Mo(N_2)_2(dppe)_2 + 2H_2$$
$$WH_2(\eta\text{-}C_5H_5)_2 + C_6H_6 \rightarrow WH(Ph)(\eta\text{-}C_5H_5)_2 + H_2$$

In some cases elimination reactions may occur intermolecularly:

$$2CoH(CO)_4 \rightarrow Co_2(CO)_8 + H_2$$

Thus, elimination of methane from cis-OsH(Me)(CO)$_4$ proceeds intramolecularly in the presence of a ligand L to afford Os(CO)$_4$L, and intermolecularly in the absence of added ligand to result in a dinuclear complex (CO)$_4$HOsOs(Me)(CO)$_4$.

2.3 Role of Rhodium Hydride Complexes in the Catalytic Hydrogenation of Olefins

The striking advances in borane chemistry during the period of 1940–1960 led to the successful development of the new field of "metal hydride chemistry" within the past 50 years. Especially transition-metal hydride complexes not only aroused academic interest but also established an exceptional position as participants in many important homogeneous catalytic reactions[76–82] such as C=C and C=O hydrogenation, C—H activation, and hydroformylation. For example, $HCo(CO)_4$ and $HCo(CO)_3$ which are produced by hydrogenolysis of $Co_2(CO)_8$:

$$H_2 + Co_2(CO)_8 \rightarrow 2\,HCo(CO)_4$$
$$HCo(CO)_4 \rightarrow HCo(CO)_3 + CO$$

are believed to be the active catalysts in hydroformylation of olefins.[83–86] In this section we briefly describe the mechanisms of catalytic hydrogenation of olefins in which rhodium hydride complexes are involved as crucial intermediates.[87, 88]

2.3.1 Homogeneous Activation of Dihydrogen by Metal Complexes in Solution[89–91]

Molecular hydrogen (dihydrogen) is cleaved by various transition-metal ions and complexes in solution, with the formation of a hydride metal complex which is generally sufficiently reactive to sustain a catalytic cycle.

2.3.1.1 Three Modes of H_2 Activation

There are three mechanisms by which such cleavage of H_2 can occur:

heterolytic splitting:

$$Cu^{2+} + H_2 \rightarrow CuH^+ + H^+ \tag{1a}$$

$$[RuCl_6]^{3-} + H_2 \rightarrow [RuHCl_5]^{3-} + H^+ + Cl^- \tag{1b}$$

homolytic splitting:

$$2\,Ag^+ + H_2 \rightarrow 2\,AgH^+ \tag{2a}$$

$$2\,[Co(CN)_5]^{3-} + H_2 \rightarrow 2\,[CoH(CN)_5]^{3-} \tag{2b}$$

dihydride formation:

$$IrCl(CO)(PPh_3)_2 + H_2 \rightarrow Ir(H)_2Cl(CO)(PPh_3)_2 \tag{3}$$

In the first of the above mechanisms, H_2 is cleaved heterolytically:

$$H_2 \rightarrow H^+ + H^-$$

and the hydride anion is accepted as a ligand, replacing one of the original ligands in the catalyst complex. In this case, the formal oxidation number of the metal atom is unchanged.

The heterolytic H_2-cleavage reaction of Eq. (1) is reversible. Hence, Cu(II) and Ru(III) can catalyze the isotopic exchange between hydrogen gas and water through forward and backward processes of Eq. (1):

$$D_2 + H_2O \rightarrow HD + HDO$$

Reaction (1a) is followed by

$$CuH^+ + Cu^{2+} \rightarrow 2Cu^+ + H^+$$

and Cu(I) can be reoxidized by substrates such as Cr(VI):

$$3Cu^+ + Cr(VI) \rightarrow 3Cu^{2+} + Cr(III)$$

Thus, Cu(II) can catalyze the oxidation of H_2 by Cr(VI) and other inorganic as well as organic oxidants.

Oxidation of H_2 by Cr(VI) is also catalyzed by Ag(I). This reaction obeys the following two-term rate law:

$$-\frac{d[H_2]}{dt} = \frac{k_1 k_2 [H_2][Ag^+]^2}{k_{-1}[H^+] + k_2[Ag^+]} + k[H_2][Ag^+]^2$$

suggesting that the overall reaction proceeds along two pathways. The first term is of the same function as the rate formula observed for the Cu(II) catalysis and attributed to the process involving the heterolytic splitting of H_2 as the rate-determining step. The second term was identified with a procedure involving the homolytic splitting of H_2 as shown by Eq. (2a).

An aqueous solution containing $CoCl_2$ and more than five equivalents of KCN absorbs H_2. This is also a third-order reaction effecting the homolytic splitting of H_2 as is shown by Eq. (2b). The $[Co(CN)_5]^{3-}$ ion serves as a homogeneous catalyst for the isotopic exchange of D_2 with H_2O, the oxidation of H_2 by substrates such as H_2O_2 and $[Fe(CN)_6]^{3-}$, and the hydrogenation of certain conjugated olefins.

2.3.1.2 Dihydride Formation by Oxidative Addition of H_2

A solution of the so-called Vaska's complex: *trans*-IrCl(CO)(PPh$_3$)$_2$ in organic solvents such as benzene, toluene, and chloroform absorbs hydrogen gas. Dihydrogen is homolytically cleaved and both hydrogen atoms are accepted as ligands in the coordination sphere of iridium without displacement of ancillary ligands [Eq. (3)]. The Ir—H bond is essentially covalent, but the shared electrons must be assigned to the more electronegative H atom in defining the oxidation numbers of Ir and H atoms. Then, both the oxidation number and coordination number of Ir increased upon reaction from 1 to 3 and 4 to 6, respectively. This kind of reaction is hence called *oxidative addition*.

Equation (3) is reversible, argon readily repelling hydrogen to recover the original Ir(I) complex. The reverse reaction causes a decrease in both the oxidation number and coordination number of the metal, and is called *reductive elimination*. The equilibrium constant of Eq. (3) is reported to be $K = 1.5 \times 10^4$ dm^3 mol^{-1} in benzene at 30°C.

The product of Eq. (3) is considered to have an octahedral structure with the *cis*-(H)$_2$-*trans*-(P)$_2$ arrangement based on IR and NMR data; crystals suitable for X-ray analysis have not been obtained. Therefore, a triangular (side-on) transition state as depicted below is proposed for Eq. (3):[92, 93]

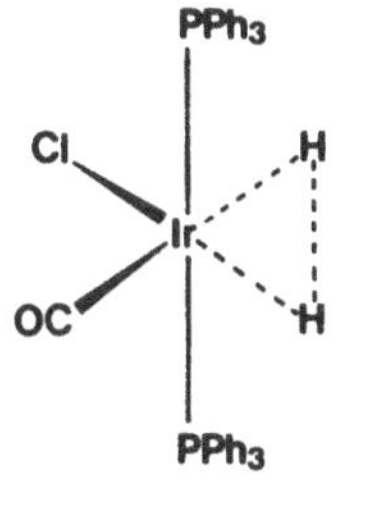

2.3.1.3 Dihydrogen Complex

Interaction of a dihydrogen molecule with Ir(I) may be prerequisit for the transition state. Complexes of Cr(0), Mo(0), Mo(II), W(0), Fe(0), Fe(II), Ru(II), Co(0), and Ir(III) containing H_2 as a η^2 (side-on bonded) ligand have been characterized very recently. Among these *trans*-W(H$_2$)(CO)$_3\{$P(*i*-Pr)$_3\}_2$[94, 95] and *trans*-[FeH(H$_2$)(dppe)$_2$]BF$_4$[96] were successfully subjected to structural analysis at room temperature.

In toluene solution $W(CO)_3\{P(i\text{-}Pr)_3\}$[95] reacted with hydrogen (1 atm) readily and cleanly, but the isolated yield of *mer-trans-*$W(H_2)(CO)_3\{P(i\text{-}Pr)_3\}_2$ was low owing to its high solubility in hydrocarbon solvents. The dihydrogen ligand is extremely labile, and storage and handling of the complex under an H_2-enriched atmosphere is necessary. Suitable single crystals were subjected to X-ray and neutron diffraction analyses at $-100(5)°C$ and room temperature, respectively.[94]

The latter Fe(II) complex of dihydrogen was prepared by the reaction of a THF or benzene solution of $Fe(H)_2(dppe)_2$ under hydrogen at 22 °C with approximately one equivalent of $HBF_4 \cdot Et_2O$. Even in the solid state, $[FeH(H_2)(dppe)_2]BF_4$ reacts slowly with nitrogen to give *trans-*$[FeH(N_2)(dppe)]BF_4$, and hence it must be stored under hydrogen or argon.[96]

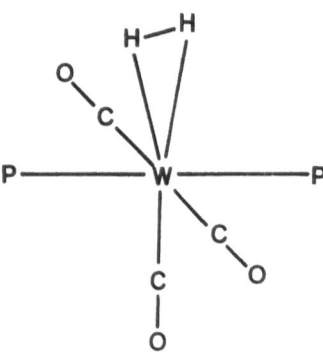

Fig. 14. Geometry of the *trans-*$W(H_2)(CO)_3\{P(i\text{-}Pr)_3\}_2$ molecule. The positions of the two hydrogen atoms shown were derived from neutron data at room temperature, while heavy-atom positions were determined by X-ray at $-100(5)°C$.[94]

Figure 14 shows the molecular structure of the tungsten complex. The geometry around the metal is an octahedron. The dihydrogen ligand is symmetrically coordinated in an η^2 mode with average W—H distances of 1.99(23) Å (X-ray) and 1.75 Å (neutron). The H—H separation is 0.75(16) Å (X-ray) and 0.84 Å (neutron), the ligand axis being approximately parallel to the *trans* P—P direction.[94]

Crystals of *trans-*$[FeH(H_2)(dppe)_2]BF_4 \cdot (THF)_2(Et_2O)$ obtained from THF solution was subjected to X-ray analysis.[96] The η^2-dihydrogen ligand is symmetrically coordinated with Fe—H distances of 1.53(8) and 1.55(7) Å, slightly longer than the terminal Fe—H distance of 1.28(8) Å. The H—H separation of 0.89(11) Å may be compared with that of 0.75(16) Å (X-ray) in the above tungsten complex, both being longer than 0.74 Å for free hydrogen gas.[96]

2.3.2 Reaction Pathway for the Hydrogenation of Olefins Catalyzed by $RhCl(PPh_3)_3$[97]

In 1966 G. Wilkinson and his coworkers[98] prepared $RhCl(PPh_3)_3$, which is a strongly active catalyst for the rapid hydrogenation of $C=C$ and $C\equiv C$ linkage at ca. 1 atm pressure of H_2 and room temperature. The overall reaction proceeds in two steps, hydrogenation of the complex and the reaction of the dihydride complex with unsaturated compounds (Scheme I).

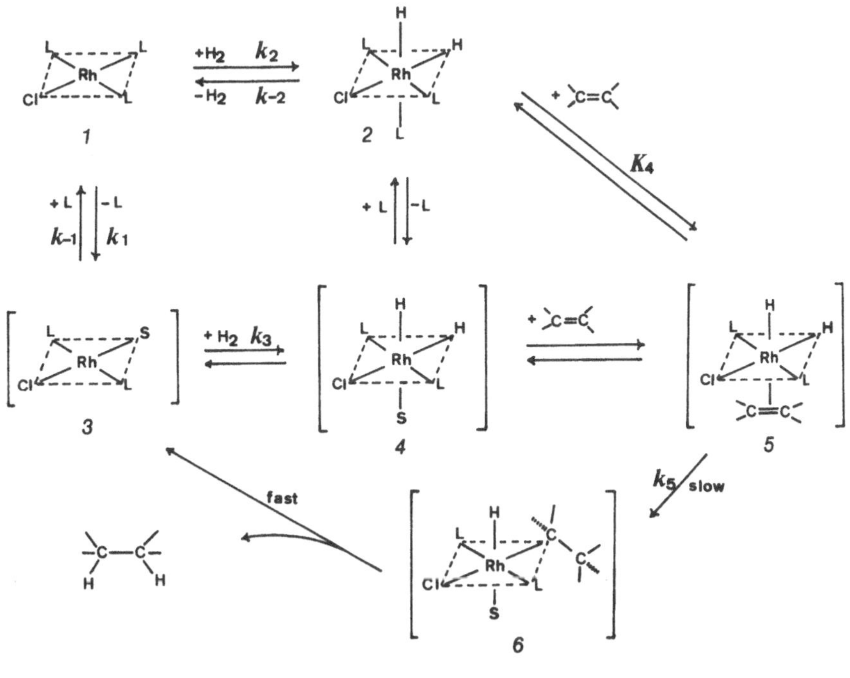

Scheme I. Reaction pathway proposed for the hydrogenation of olefins catalyzed by $RhCl(PPh_3)_3$. Here L stands for PPh_3 and the bracketed species are not isolated.[97]

When hydrogen gas is added to a solution of $RhCl(PPh_3)_3$ in dichloromethane, the UV absorption spectrum changes progressively with the amount of H_2 added, exhibiting a sharp isosbestic point at 360 nm. This is caused by the oxidative addition of H_2 to the Rh(I) complex:

$$RhCl(PPh_3)_3 + H_2 \rightarrow Rh(H)_2Cl(PPh_3)_3 \qquad (4)$$

$$\quad\; 1 \qquad\qquad\qquad\qquad\qquad 2$$

This reaction is reversible, K being $(4.7 \pm 0.8) \times 10^3 \, \mathrm{dm^3 \, mol^{-1}}$ in CH_2Cl_2 at 25°C.[99]

Kinetic analysis revealed that although the degree of dissociation of 1 to form 3 is very small $(K = (7 \pm 1) \times 10^{-5} \, \mathrm{mol \, dm^{-1}}$ in benzene at 25°C),[100] 3 reacts with H_2 about 10^4 times faster than 1 does, and Eq. (4) proceeds actually via 3 and 4.[101]

When a benzene solution of 1 is saturated with H_2 to convert the complex to 2 completely in the presence of added PPh_3, and then a large excess of olefin such as cyclohexene is added to this solution, the following hydrogenation reaction occurs quantitatively:

$$Rh(H)_2 Cl(PPh_3)_3 + C_6H_{10} \rightarrow RhCl(PPh_3)_3 + C_6H_{12} \qquad (5)$$
$$\qquad\quad 2 \qquad\qquad\qquad\qquad\qquad\quad 1$$

The rate of the reaction of Eq. (5) determined by a spectrophotometric method obeys the pseuco-first-order rate law:

$$d[1]/dt = -d[2]/dt = k_{obs}[2]$$

$$k_{obs} = \frac{k_5 K_4 [C_6H_{10}]}{[PPh_3] + K_4 [C_6H_{10}]}$$

suggesting that the reaction proceeds via a ligand substitution equilibrium followed by the rate-determining insertion of olefin into the Rh-H bond (actually migration of a hydride ligand onto the olefinic carbon):

$$RhCl(H)_2(PPh_3)_3 + C_6H_{10} \xrightarrow{\;K_4\;} RhCl(H)_2(C_6H_{10})(PPh_3)_2 + PPh_3$$
$$\qquad\qquad 2 \qquad\qquad\qquad\qquad\qquad\qquad\qquad\quad 5$$

$$5 \xrightarrow[\text{slow}]{k_5} 6 \xrightarrow[\text{fast}]{PPh_3} RhCl(PPh_3)_3 + C_6H_{12}$$
$$\qquad\qquad\qquad\qquad\qquad\qquad\quad 1$$

Here, $K_4 = (3.4 \pm 0.6) \times 10^{-4}$ and $k_5 = 0.20 \pm 0.04 \, \mathrm{s^{-1}}$ in benzene at 25°C. Reductive elimination of the dihydrogenated product from 6 to reproduce 3 is rapid.[102]

Thus, activation of both the olefin and H_2 by coordination (formation of 5) is essential for the catalysis by the Wilkinson complex, and in the absence of excessive PPh_3 the catalytic cycle seems to work in the sequence $3 \rightarrow 4 \rightarrow 5 \rightarrow 6 \rightarrow 3$.

2.3.3 Hydrogenation of Olefins Catalyzed by Cationic Rh(I) Complexes, $[Rh(PP)(S)_2]^+$ [103]

2.3.3.1 Formation of Cationic Rh(I) Complexes

A cyclic diene ligand such as norbornadiene involved in a Rh(I) bis-phosphine complex is easily hydrogenated by H_2 and removed from the coordination sphere. When the reaction is conducted in a donor solvent such as MeOH, MeCN, and THF, the vacated coordination sites are occupied by solvent molecules. The dominant products of the reaction, however, depend on the nature of the phosphine ligands. A complex having two unidentate phosphines requires three molecules of H_2, affording an octahedral dihydrido Rh(III) complex [Eq. (6)]. On the other hand, a complex having a chelating diphosphine absorbs two molecules of H_2 to yield a square-planar Rh(I) complex [Eq. (7)]:

$$ (6) $$

$$ (7) $$

The difference between these two reactions may be caused by the *trans* influences of the hydride and phosphine ligands. Usually, phosphines avoid becoming *trans* to hydrides when possible. Therefore, complex 7 resists the reaction with another molecule of H_2, since the oxidative addition is *cis*, requiring at least one site *trans* to P.

2.3.3.2 Mechanism of Olefin Hydrogenation

The mechanism of hydrogenation involving the cationic diphosphine catalysts has been elucidated by J. Halpern and others, and is shown in Scheme 2. An olefin complex 8 is in equilibrium with catalyst 7. Values of

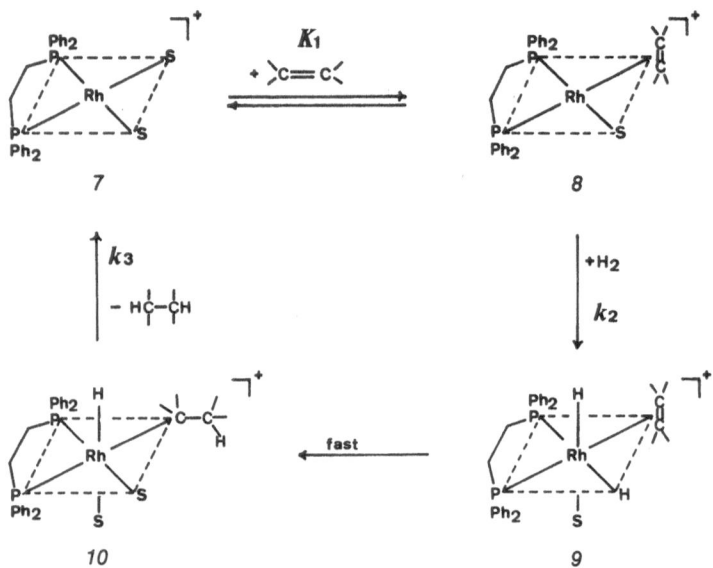

Scheme 2. Reaction pathway proposed for the hydrogenation of olefins catalyzed by [Rh(dppe)(S)$_2$]$^+$. Here S denotes a coordinated solvent molecule.[103]

the equilibrium constant K_1 were determined spectrophotometrically in
methanol at 25°C resulting in the following sequence:

unsat. compd.	1-hexene	benzene	styrene	toluene
K_1/dm^3 mol^{-1}	2	18	20	97

α-acetamidocinnamic acid

8×10^3

The affinity of unsaturated compounds to 7 strongly depends on their
structures. Coordination of aromatic rings was confirmed by isolation
and X-ray structure determination of the dimer [Rh$_2$(dppe)$_2$][BF$_4$]$_2$
which was crystallized from methanol.[104] The extraordinarily large
value of K_1 for α-acetamidocinnamic acid and its esters is caused by
coordination through the amido oxygen as well as the C=C group.
Thus, the molecular structure of [Rh(dppe)(MAC)] BF$_4$ determined by
X-ray analysis confirmed this coordination mode, where MAC is methyl
(Z)-α-acetamidocinnamate.[105]

Oxidative addition of H$_2$ (k_2) is the rate-determining step at room
temperature. The dihydride complex 9 is not stable as mentioned above,
rapidly forming 10 by migratory insertion. Subsequent reductive

elimination of the product from 10 to reproduce catalyst 7 is also rapid at ambient temperatures. At lower temperatures (e.g. at $-78\,°C$), however, the final step (k_3) becomes the rate-determining step.

In the case of the above MAC complex, for example, $K_1 = 5.3 \times 10^3 \ dm^3 \ mol^{-1}$ and $k_2 = 1.0 \times 10^2 \ dm^3 \ mol^{-1} \ s^{-1}$ at $25\,°C$, $\Delta H_2^{\ *} = 6.3 \ kcal \ mol^{-1}$, $\Delta S_2^{\ *} = -28 \ cal \ deg^{-1} \ mol^{-1}$, whereas $k_2 = 0.13 \ dm^3 \ mol^{-1} \ s^{-1}$ at $-78\,°C$ and $k_3 = 6.0 \times 10^{-4} \ s^{-1}$ at $-56.4\,°C$, $\Delta H_3^{\ *} = 17.0 \ kcal \ mol^{-1}$, $\Delta S_3^{\ *} = 6.0 \ cal \ deg^{-1} \ mol^{-1}$. Therefore, the rates of the oxidative addition of $H_2 (k_2)$ and the reductive elimination from the alkyl complex (k_3) are comparable at about $-40\,°C$ under 1 bar of H_2, and the former reaction is rate determining at higher temperatures and the latter at lower temperatures. Thus, the intermediate 10 was intercepted at $-78\,°C$ and characterized by ^1H, ^{13}C, and ^{31}P NMR spectroscopy to reveal clearly that H transfer during the migratory insertion step occurs to the β-carbon atom of the C=C bond, while the α-carbon atom becomes bonded to the rhodium.[106]

Thus, the mechanisms of olefin hydrogenation are different for the catalyst systems. The Wilkinson complex accepts H_2 prior to olefin, whereas the cationic Rh(I) catalyst forms the olefin complex first, which reacts with H_2 subsequently. In either case the dihydride complex is the key intermediate, and two hydride ligands add successively to the olefin molecule situated at the adjacent coordination site.

2.3.3.3 Asymmetric Hydrogenation[107–110]

By employing a chiral diphosphine[109] such as (R,R)-1,2-bis{(o-methoxy-phenyl)phenylphosphino}ethane ((R,R)-DIPAMP, 11), (S,S)-2,3-bis(di-phenylphosphino)butane ((S,S)-CHIRAPHOS, 12), and (R,R)-trans-4,5-bis(diphenylphosphinomethyl)-2,2-dimethyldioxolan ((R,R)-DIOP, 13) as a chelating ligand in catalyst 7, asymmetric hydrogenation of prochiral olefinic substrates such as α-acylaminocinnamates has become possible.

11   12   13

For example, (Z)-α-benzylaminocinnamic acid is hydrogenated by [Rh((S,S)-CHIRAPHOS)(S)$_2$]$^+$ in THF to afford R-phenylalanine derivative in 99% ee.[111] The kinetic and mechanistic aspects of the asymmetric hydrogenation have been extensively discussed elsewhere.[110]

3 Diatomic Ligands

Besides the anionic ligands such as OH^- and CN^-, there exist various diatomic neutral ligands. Particularly metal complexes containing N_2, O_2, NO, and CO have been extensively investigated owing not only to the structural interest but also to the important roles they play in the synthetic, biological, and environmental chemistry. Among them carbonyl and dinitrogen complexes are taken up in this chapter, since they have been investigated most actively in recent years.

3.1 Coordination Modes for Carbon Monoxide[112]

Coordination modes for CO in metal complexes are summarized in Fig. 15. The first row shows the well-known terminal, edge-bridging,[113] and face-bridging carbonyl arrays. In either case, CO is bonded exclusively through carbon. No example has been reported of CO being bonded solely through oxygen.

3.1.1 Terminal Bonding and μ(C) Bridging

The terminal M—C—O linkage is essentially linear, the bond angles around C ranging from 165 to 175°. In most of the edge-bridging cases the C—O vectors are almost perpendicular to the M—M bond and the M—C—M angles are in the region of 77–90°. When CO bridges two different kind of metal atoms, or two same kind of metal atoms which are coordinated with ancillary ligands of remarkably different natures, the bridging geometry may be asymmetric. The term "semibridging" has been proposed for the case where the two M—C bond lengths and M—C—O angles differ more than 0.3 Å and 20°, respectively.[113]

The vibrational spectra, particularly the infrared spectra are useful in deducing the coordination mode of the carbonyl ligand in metal complexes. The terminal CO generally absorbs at high frequencies (1850–2125 cm^{-1}), while the edge-bridging and face-bridging CO

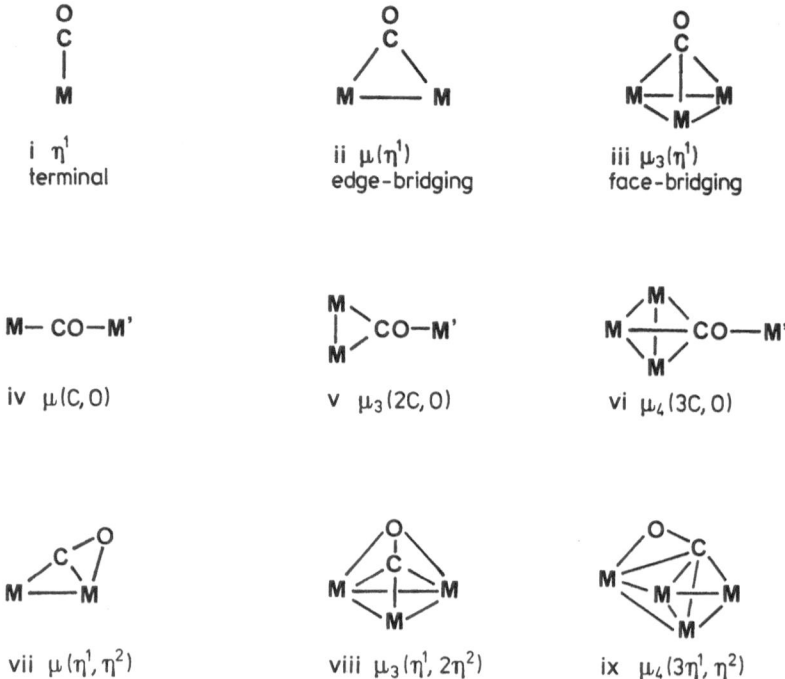

i η¹
terminal

ii μ(η¹)
edge-bridging

iii μ₃(η¹)
face-bridging

iv μ(C, O)

v μ₃(2C, O)

vi μ₄(3C, O)

vii μ(η¹, η²)

viii μ₃(η¹, 2η²)

ix μ₄(3η¹, η²)

Fig. 15. Coordination modes for CO in metal complexes.

groups absorb in the ranges of 1700–1860 cm^{-1} and 1620–1730 cm^{-1}, respectively. The frequencies are progressively lower with decreasing bond order of CO compared with 2143 cm^{-1} for free gaseous molecule. It must be noticed that the ν(CO) frequency is affected by the nature of ancillary ligands and the electric charge on the complex.

All other entries in Fig. 15 represent some degree of oxygen bonding as well.

3.1.2 M$_x$—CO—M′ Bridging

Compounds (iv)–(vi) in the second row in Fig. 15 contain a CO ligand of which the carbon atom is terminally bonded to a metal atom or bridging two or three metal atoms, and the oxygen atom is further bonded to another metal atom M′.

As was discussed on p. 2, the bonding mechanism of CO is synergic and π-back-donation from metal to CO is slightly more than offsetting the σ-donation, with the result that the metal atoms have a slightly positive charge and the oxygen atom has a slightly negative charge. For similar compounds, the carbonyl basicity increases with the number of metal atoms bonded to a single CO in the sequence: terminal CO < edge-bridging CO < face-bridging CO. Hence, the bridging CO affords more stable adducts than the terminal CO.

Upon formation of these adducts, the bond order and hence the stretching frequencies of the bridged CO are greatly reduced, while the M—C bond is strengthened in turn. Figures 16, 17, and 18 display recent examples.

3.1.2.1 M—CO—M′ Systems

Crystals of $(\eta\text{-}C_5H_5)_2ZrMe(\mu\text{-}OC)Mo(CO)_2(\eta\text{-}C_5H_5)$ were obtained from the solid-state reaction of $(\eta\text{-}C_5H_5)_2ZrMe_2$ with $HMo(CO)_3(\eta\text{-}C_5H_5)$ in the presence of a small amount of THF, and subjected to X-ray analysis. The molecular structure shown in Fig. 16 exemplifies the shorter Mo—C(1) (1.847(5) Å) and longer C(1)—O(1) (1.236(5) Å) distances for the bridged CO in comparison with Mo—C(2) (1.917(12) Å) and C(2)—O(2) (1.165(16) Å) for a non-bridged CO. The $v(CO)$ frequency of 1545 cm^{-1} assigned to the bridged CO is remarkably lower than 1948 and 1863 cm^{-1} assigned to the terminal CO. The Mo—C(1)—O(1) bond is linear (178.2(6)°), but the C(1)—O(1)—Zr linkage is non-linear with the angle 145.5(4)°.[114]

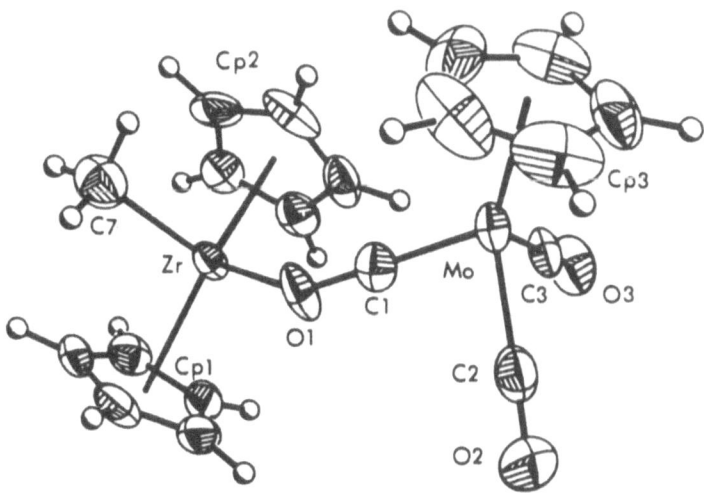

Fig. 16. Molecular structure of $(\eta\text{-}C_5H_5)_2ZrMe(\mu\text{-}OC)Mo(CO)_2(\eta\text{-}C_5H_5)$.[114]

3.1.2.2 M$_2$—CO—M′ Systems

The first described case of a polynuclear metal complex containing a $\mu(C, O)$-bridged CO group was $(\eta\text{-}C_5H_5)_2Fe_2(CO)_4 \cdot 2\,AlEt_3$ which

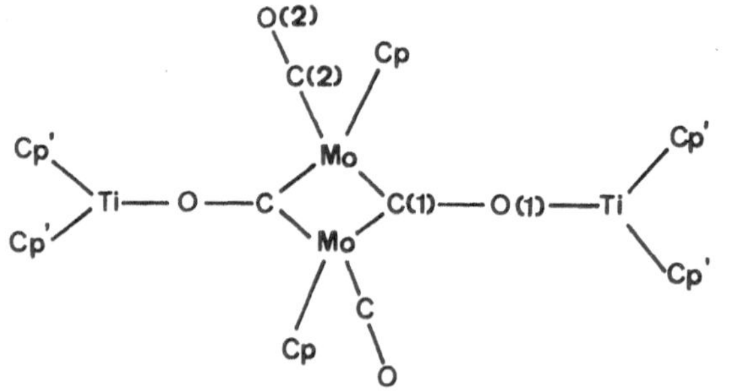

Fig. 17. Molecular structure of
$[(\eta\text{-}C_5Me_5)_2Ti]_2(\mu\text{-}OC)_2[(\eta\text{-}C_5H_5)Mo(CO)]_2$.[117]

Shriver et al. obtained in 1969 from the simple addition reaction between
$(\eta\text{-}C_5H_5)_2Fe_2(CO)_4$ and slightly excess Al_2Et_6 in deaerated
benzene.[115, 116] Since then, a number of similar compounds have been
prepared. Figure 17 illustrates the structure of $[(\eta\text{-}C_5Me_5)_2Ti]_2(\mu\text{-}$
$OC)_2[(\eta\text{-}C_5H_5)Mo(CO)]_2$ as a recent example, which was prepared by
reaction of $(\eta\text{-}C_5Me_5)_2Ti(neo\text{-}C_5H_{11})$ with $[(\eta\text{-}C_5H_5)Mo(CO)_2]_2$ in
benzene at room temperature. The molecule is centrosymmetric and the
plane formed by the two (C_5Me_5)-ring centroids, Ti, and O is nearly
perpendicular to the plane formed by the two Ti atoms and the atoms of
the Mo_2—$(\mu\text{-}CO)_2$ core: the dihedral angle is 93.5°. The bridging $C(1)$—
$O(1)$ distance $(1.271(7)\text{Å})$ is longer than that of the terminal $C(2)$—$O(2)$
$(1.163(11)$ Å). The IR bands at 1351 and 1872 cm^{-1} are assigned to their
$\nu(CO)$ absorptions, respectively.[117]

Hitherto, many compounds containing cyclopentadienyl complexes
of the early transition and lanthanoid metals (e. g. $(\eta\text{-}C_5H_5)_2M$ with
$M = Ti$, Zr, Hf, and Yb) coordinated to metal carbonyls have been
structurally characterized. In these complexes, the $(\eta\text{-}C_5H_5)_2M$ unit is
coordinated with the carbonyl oxygen, exhibiting the $\mu(C, O)$ mode, and
all show a non-linear M—O—C arrangement as is exemplified by
Fig. 16. In the present complex, the $C(1)$—$O(1)$—Ti linkage is linear with
a bond angle of 177.9(4)°. This is the first case of a linear C—O—M chain
and was qualitatively explained based on the MO consideration
deducing that both the Ti—$O(1)$ and $C(1)$—$O(1)$ bonds are of approximate bond order two.[117]

3.1.2.3 M₃—CO—M′ Systems

Recently, a novel CO linkage was found in the molecular structure of
$Os_6(CO)_{17}(py)_2$ (Fig. 18) which was obtained as a minor product from
the reaction of $Os_6(CO)_{18}$ with excess pyridine. The molecule has a

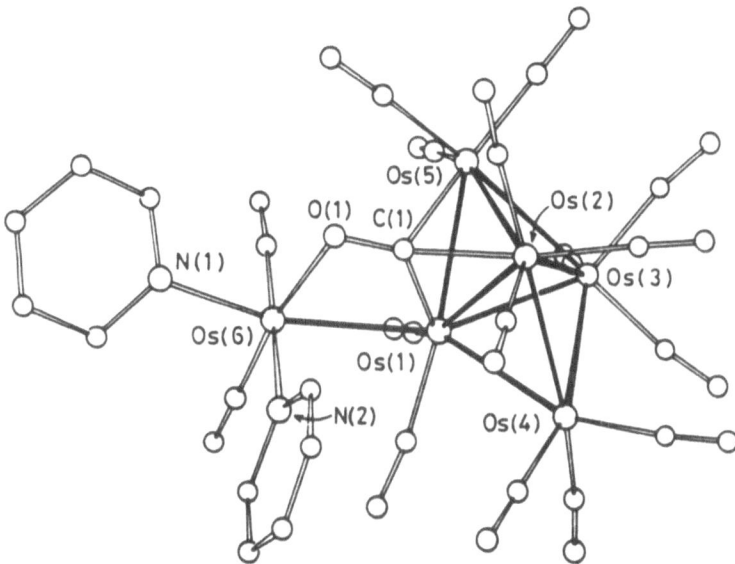

Fig. 18. Molecular structure of $Os_6(CO)_{17}(py)_2$.[118]

metal core consisting of a trigonal-bipyramid arrangement of five Os atoms with the sixth, Os(6), bonded to an equatorial atom, Os(1), in a 'spike' arrangement. Sixteen CO groups are terminal and the remaining C(1)—O(1) is face-bridging: Os(1)—C(1) 1.98(2), Os(2)—C(1) 2.18(2), and Os(5)—C(1) 2.29(2) Å. The O(1) atom occupies an octahedral coordination site of Os(6) at a bonding distance of 2.139(13) Å. The C(1)—O(1) distance is recorded as 1.28(2) Å, but unfortunately a low-frequency v(CO) band assignable to the bridging CO is not observed. The C(1)—O(1)—Os(6) angle is 95(1)° and C(1) is far from Os(6) (2.58(2) Å) with no bonding interaction.[118]

3.1.3 CO Bridge Involving the η^2(side-on) Linkage

The bottom row of Fig. 15 illustrates those compounds which involve the η^2(side-on) linkage of CO. This linkage mode of CO is supposed as a prerequisite for the cleavage of CO on surfaces of metals which are employed as catalysts in Fischer-Tropsch synthesis. Although the experimental evidences supporting this idea are not satisfactory in the field of heterogeneous catalysis, a substantial number of structurally well-characterized examples have recently been found in the field of coordination chemistry.

3.1.3.1 $\mu(\eta^1, \eta^2)$ Mode

The first dinuclear type-(vii) complex described is $Mn_2(CO)_5(dppm)_2$ reported in 1975.[119-121] This compound was prepared by the reaction of $Mn_2(CO)_{10}$ with a twice molar amount of dppm in refluxing n-decane or xylene under nitrogen. The X-ray molecular structure of this red diamagnetic complex is depicted in Fig. 19. The two Mn atoms are held together at a bonding distance of 2.934(6) Å by virtue of two bridging dppm ligands. The five carbonyl groups lie approximately in a plane which passes through the metal atoms and is nearly perpendicular to the Mn_2P_4 plane.

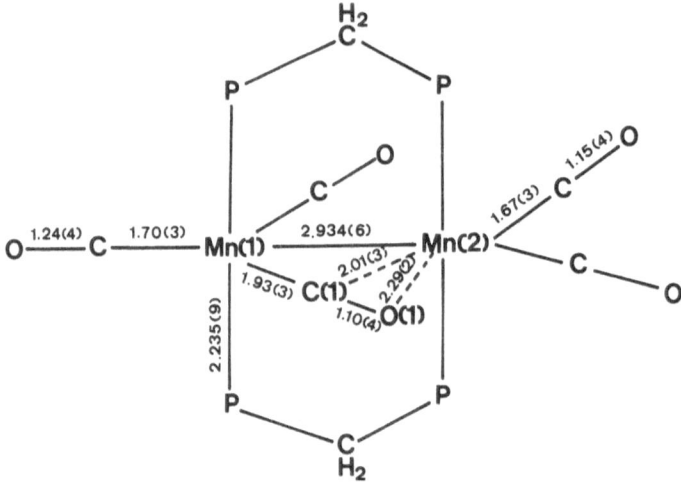

Fig. 19. The atomic arrangement in $Mn_2(CO)_5(dppm)_2$.[120]

Four of the CO molecules are normal as terminal ligands, but the fifth is unique. The Mn(1)—C(1)—O(1) array is linear (173(3)°) with the CO tipped toward Mn(2). The Mn(2)—C(1) (2.01(3) Å) and Mn(2)—O(1) (2.29(2) Å) distances indicate the η^2 bonding of C(1)—O(1) to Mn(2). An IR band at 1645 cm^{-1} is assigned to the $v(CO)$ vibration of this unique CO ligand. Thus, it seems reasonable to consider the $\mu(\eta^1, \eta^2)$-CO as a four-electron donor, affording two electrons to Mn(1) by the usual terminal (η^1) bonding and two to Mn(2) by the side-on (η^2) bonding. This bonding scheme brings the electron count on each Mn to the expected 18.[121] A number of additional binuclear complexes containing a similar $\mu(\eta^1, \eta^2)$-CO linkage have been reported including heteronuclear complexes such as $(\eta\text{-}C_5H_5)Co(CO)_2Zr(\eta\text{-}C_5Me_5)_2$ and $(\eta\text{-}C_5H_5)Mo(CO)_2(\eta^2\text{-}CH_3CO)Zr(\eta\text{-}C_5H_5)_2$.[112]

3.1.3.2 $\mu_3(\eta^1, 2\eta^2)$ Mode

The X-Ray crystal structure of $(\eta\text{-}C_5H_5)_3Nb_3(CO)_7$ is shown in Fig. 20; this is the only example of a type-(viii) complex in Fig. 15. This was the first described Nb carbonyl cluster obtained from the photochemical trimerization of $(\eta\text{-}C_5H_5)Nb(CO)_4$ in hexane at 18 °C.[122]

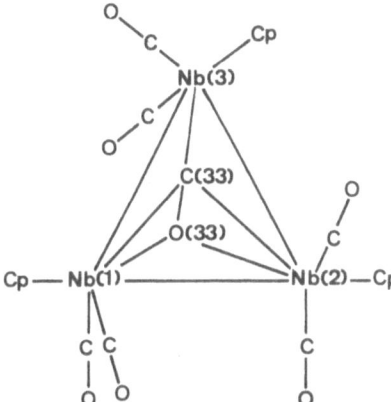

Fig. 20. Molecular structure of $(\eta\text{-}C_5H_5)_3Nb_3(CO)_7$ viewed from the top of the $\mu_3(\eta^1,2\eta^2)$-CO ligand.[122]

Each Nb atom constituting a nearly equilateral triangle is coordinated with a cyclopentadienyl ring and two terminal CO ligands. The remaining CO acts as a $\mu_3(\eta^1, 2\eta^2)$ ligand with a dramatically length-ened C—O distance (1.303(14) Å) and a long-wave shift of the $\nu(CO)$ absorption (1330 cm^{-1}). Conversely, the Nb(3)—C(33) distance is the shortest (1.966(12) Å). Both of the C(33) and O(33) atoms lie near Nb(1) and Nb(2) almost in the bonding range, e. g. Nb(1)—C(33) = 2.278(14) Å and Nb(1)—O(33) = 2.212(10) Å. The Nb(1)—C(33)—Nb(2) and Nb(1)—O(33)—Nb(2) angles are both close to 90° (85.0(4) and 86.3(3)°, respectively), and C(33)—O(33) appears to be bonded to Nb(1) and Nb(2) by donating two pairs of π-electrons similarly to the acetylene molecule in $Co_2(CO)_6(PhC\equiv CPh)$. Thus, the unique CO ligand functions as a six-electron donor toward a total of three metal centers, satisfying the 18-electron rule.[122]

3.1.3.3 $\mu_4(3\eta^1, \eta^2)$ Mode

Two cases of a type-(ix) complex (Fig. 15) have been found so far. One is $Co_2Mo_2(\mu\text{-}CO)_3(\mu_4\text{-}CO)(\eta\text{-}C_5H_5)_2(\eta\text{-}C_5Me_5)_2$[123] and the other is anion in $[Me_3(PhCH_2)N][Fe_4(CO)_{13}H]$.[124]

The former Co_2Mo_2 complex was prepared by ultraviolet irradia-tion of an equimolar mixture of $(\eta\text{-}C_5Me_5)Co(C_2H_4)_2$ and $(\eta\text{-}C_5H_5)_2Mo_2(CO)_4$ in toluene at 50–60 °C. Its molecular structure

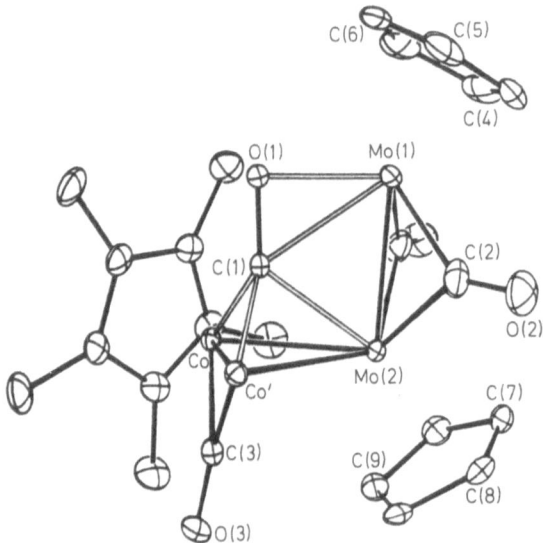

Fig. 21. Molecular structure of $Co_2Mo_2(\mu\text{-CO})_3(\mu_4\text{-CO})(\eta\text{-}C_5H_5)_2(\eta\text{-}C_5Me_5)_2$. The C_5Me_5 ring attached to Co' is omitted for clarity.[123]

established by X-ray diffraction study is shown in Fig. 21. The metal core is open, with an isoceles triangular $Co_2Mo(2)$ fragment attached to a second molybdenum by a Mo=Mo double bond (2.574(1) Å) nearly perpendicular to the trimetal plane. Of the four CO ligands, C(1)O(1) is unique, the vector being almost perpendicular (at 88.5°) to the $Co_2Mo(2)$ plane. The Mo(1)—O(1) distance (2.088(2) Å) is smaller than that of Mo(1)—C(1) (2.404(3) Å), and the η^2-bonding of C(1)—O(1) to Mo(1) is evident. The C(1)—O(1) bond is so much lengthened (1.283(3) Å), indicating that it is substantially activated.[123]

As is shown in Fig. 22, four iron atoms in $[Fe_4(CO)_{13}H]^-$ constitute a butterfly-shaped framework with a dihedral angle of 177° which

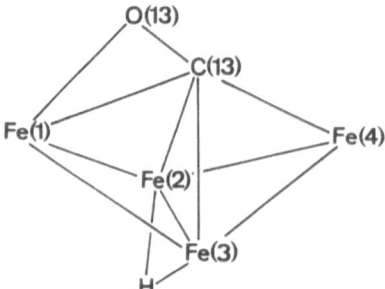

Fig. 22. Cluster framework of $[Fe_4(CO)_{13}H]^-$ in its $[Me_3(PhCH_2)N]^+$ salt, showing the $\mu_4(3\eta^1, \eta^2)$ coordination of a CO molecule; each Fe atom has three terminal CO ligands.[124]

resulted by scission of one Fe—Fe bond in the tetrahedral anion $[Fe_4(CO)_{13}]^{2-}$ upon protonation. Twelve of the thirteen CO groups are terminally bonded, three to each Fe atom, while the remaining CO is trapped among the two butterfly wings in a unique bonding situation. The Fe(4)—C(13) and C(13)—O(13) distances are 1.81(2) and 1.26(3) Å, respectively. The C(13) atom is at bonding distance from the other three Fe atoms (2.10(2)–2.17(2) Å), and O(13) is tilted toward Fe(1) to give rise to a short bonding interaction (2.00(2) Å). This CO ligand therefore behaves as a four-electron donor, allowing the cluster to meet the noble gas formalism.[124]

3.2 CO Cleavage and Reduction

In the Fischer-Tropsch (F.T.) synthesis, a mixture of CO and H_2 (synthetic gas or syngas), which is produced by burning coal and other carbonaceous materials in the presence of oxygen and steam, is converted into a wide range of hydrocarbon and oxygen-containing products.[125–131] Since the finding by Fischer and Tropsch in 1923, various supported metals such as Fe, Ru, Co, Rh, Ir, Ni, Pd, and Pt have been examined as catalysts.[127, 128]

In recent years homogeneous systems containing organometallic compounds, especially metal carbonyl clusters as catalysts have been investigated very actively with the aim of selectivity improvement and mechanism elucidation.

Of the three main mechanisms proposed for the heterogeneous F.T. synthesis of hydrocarbons, the following "carbide-methylene" mechanism is most widely used:[132,133]

Upon chemisorption on a metal catalyst the CO molecule dissociates into carbon and oxygen atoms.[132] The surface carbide thus produced is hydrogenated successively to CH_x entities, which then link up to afford the products.[133]

Isolation and characterization of intermediates are much easier for the homogeneous reactions of organometallic compounds than for similar heterogeneous reactions on solid catalysts. Various model reactions for the F.T. synthesis have been extensively studied. Thus, a number of metal complexes containing a methylene group as a terminal or bridging ligand have been prepared, and reactions of these ligands inclusive of C—C bond formation have been investigated.

As was discussed in Section 3.1, the C—O bond is lengthened in both $\mu(C, O)$ and $\mu(\eta^1, \eta^2)$ environments, suggesting an activation of the C—O bond. Cleavage of the C—O bond and production of CH_4 is effected by the use of strong acids including Brønsted acids such as HSO_3CF_3 and HCl and Lewis acids, e.g., BF_3 and $AlCl_3$. Thus, the reaction of $[Fe_4(CO)_{13}]^{2-}$ with neat HSO_3CF_3 results in degration of the cluster and formation of methane, and the steps included in this overall reaction are assumed as depicted in Scheme 3.[134]

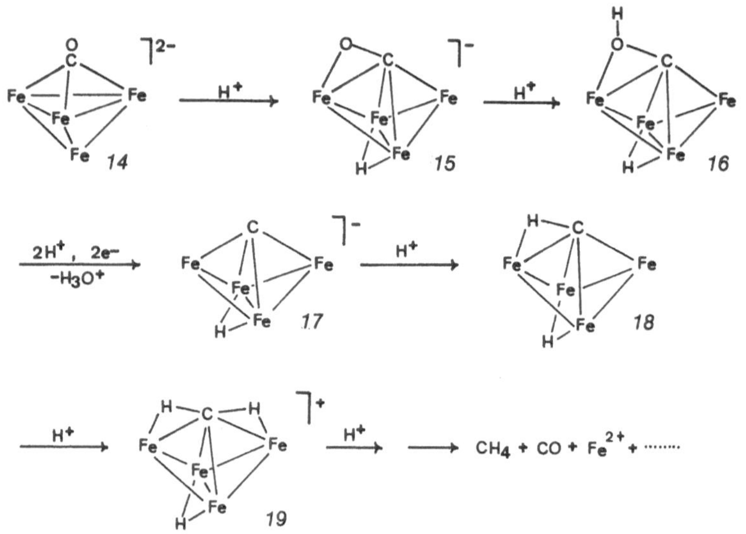

Scheme 3. Proposed steps for the reaction of $[Fe_4(CO)_{13}]^{2-}$ with strong acids such as neat HSO_3CF_3. Terminal CO ligands are omitted for clarity.[134]

Molecular structures of the starting complex *14*[135] and intermediates *15* (Fig. 22) and *17*[134] were determined by X-ray analysis, and that of *18* by neutron diffraction.[136] Intermediate *16* has not been isolated, but its methyl derivative $(\mu\text{-H})Fe_4(CO)_{12}(\eta^2\text{-COCH}_3)$ was prepared separately and its structure determined by X-ray analysis.[134] Except *14* which has a tetrahedral Fe_4 core, clusters *15–18* have the butterfly structure. Details of steps between *16* and *17* are lacking and reactions beyond *18* are speculative, but the possible role of the carbide species in the reduction of CO is manifested by these studies.

The mechanism of F.T. synthesis to afford oxygen-containing products such as alcohols, acids, and esters seems to be somewhat different
from that for the hydrocarbon production, and is being widely
investigated.[129]

3.3 Coordination Modes for Dinitrogen

The total amount of dinitrogen fixed on a global basis is estimated at
about 2.4×10^8 tons per year, and some 65% of this is pursued by
nitrogen-fixing microorganisms. While the industrial Haber process,
which accounts for about 25% of the total, is carried out at high
temperatures (300–400 °C) and pressures (350–1000 atm), the natural
biological fixation proceeds under ambient conditions.

The nitrogen-fixing enzyme nitrogenase was first isolated in 1960 and
consists of two proteins, an Fe protein at a molecular weight of around
68000 and a Mo-Fe protein with a molecular weight of 220000. An
unusual molybdenum-iron-sulfur cluster, the iron-molybdenum
cofactor "FeMo-co", is thought to provide the active site for the catalytic
reduction of dinitrogen to ammonia. Various efforts are being made to
disclose the exact structure of FeMo-co.[137]

There have been many attempts to achieve nitrogen fixation along
three other lines:[138] (1) synthesis of model clusters which should exhibit
nitrogenase activity, (2) search for transition metal systems capable of
reducing N_2 in solution by virtue of appropriate reducing agents, and (3)
activation of N_2 through ligation to transition metal atoms, leading to
nitrogen hydrides and nitrogen-carbon bonds.[139, 140]

Six coordination modes for N_2, as depicted in Fig. 23, have been
proposed so far. A dinitrogen molecule retains a pair of unshared

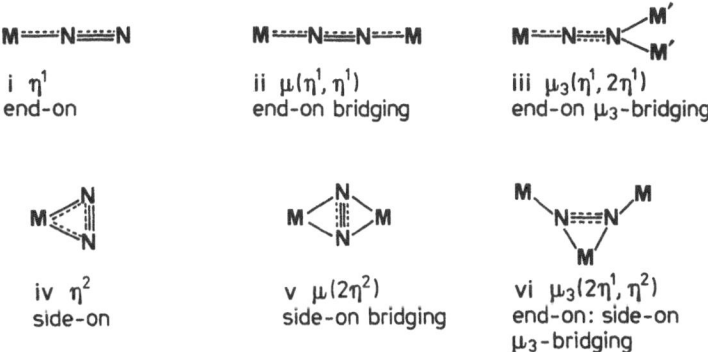

Fig. 23. Coordination modes for N_2 in metal complexes.

Table II. Examples of Metal Complexes Containing the End-on Unidentate and End-on Bridging Dinitrogen Ligands

Compound*	M—N, Å	N—N, Å	v(N—N), cm^{-1}	Ref.
N$_2$		1.0976(1)	2360	a, b
End-on unidentate				
trans-Cr(N$_2$)$_2$(dmpe)$_2$	1.874(3)	1.122(3)	1930	c
cis-Mo(N$_2$)$_2$(PMe$_3$)$_4$	1.97(1)	1.15(1)	2010, 1965	d
		1.14(1)		
trans-Mo(N$_2$)$_2$(dppe)$_2$	2.014(5)	1.118(8)	2020, 1970	e
trans-Mo(CO)(N$_2$)(dppe)$_2$	2.068(12)	1.087(18)	2110, 2080	f
cis-W(N$_2$)$_2$(PMe$_2$Ph)$_4$	1.983(3)	1.125(4)		g
	2.015(3)	1.120(4)		
trans-ReCl(N$_2$)(PMe$_2$Ph)$_4$	1.966(21)	1.055(30)	1922	h
mer-ReCl(N$_2$)(MeNC){P(OMe)$_3$}$_3$	1.980(14)	1.038(21)	2020	i
cis-Re(NHPh)(N$_2$)(PMe$_3$)$_4$	1.955(13)	1.101(18)		j
trans-[RuN$_3$(N$_2$)(en)$_2$]PF$_6$	1.894(9)	1.106(11)	2103	k
(PPh$_3$)$_2$(H)$_2$Ru(μ-H)$_4$Ru(N$_2$)—				
(PPh$_3$)$_2$	1.94(4)	1.11	2140	l
[Os(NH$_3$)$_5$(N$_2$)]Cl$_2$	1.842(13)	1.12(2)	2025, 2010	m
CoH(N$_2$)(PPh$_3$)$_3$	1.784(13)	1.102(12)	2088	n
	1.829(12)	1.123(13)		
trans-RhH(N$_2$){PPh(t-Bu)$_2$}$_2$	1.970(4)	1.074(7)	2155	o
trans-RhCl(N$_2$){P(i-Pr)$_3$}$_2$	1.885(4)	0.958(5)	2100	p
End-on bridging				
[(η-C$_5$Me$_5$)$_2$Ti]$_2$(μ-N$_2$)	2.005(10)	1.165(14)		q
	2.016(10)			
	2.013(10)	1.155(14)		
	2.033(10)			
[(η-C$_6$H$_3$Me$_3$)Mo(dmpe)]$_2$(μ-N$_2$)	2.042(4)	1.145(7)	1989	r
[(η-C$_5$H$_4$Me)Mn(CO)$_2$]$_2$(μ-N$_2$)	1.875(5)	1.118(7)	1971	s
(PMe$_2$Ph)$_4$ClRe(μ-N$_2$)MoCl$_4$(OMe)		1.18(3)	1660	t
Re—N	1.815(15)			
Mo—N	1.90(1)			
[(PM$_2$Ph)$_4$ClRe(μ-N$_2$)]$_2$MoCl$_4$		1.154(29)	1800	u
Re—N	1.888(21)			
Mo—N	1.975(20)			
[{(NH$_3$)$_5$Ru}$_2$(μ-N$_2$)][BF$_4$]$_4$	1.928(6)	1.124(15)	2050–2100	v
[MeC(CH$_2$PPh$_2$)$_3$Co]$_2$(μ-N$_2$)	1.76(1)	1.18(2)		w
[(PMe$_3$)$_3$Co(μ-N$_2$)]$_2$Mg(THF)$_4$		1.18	2068	x
Co—N	1.72			
Mg—N	2.04			
[{P(i-Pr)$_3$}$_2$HRh]$_2$(μ-N$_2$)	1.977(6)	1.134(5)	2140	y

* Solvent of crystallization is omitted.

a) Stoicheff, B. P.: Canad. J. Phys. *32*, 630 (1954); b) Nakamoto, K.: Infrared and Raman Spectra of Inorganic and Coordination Compounds p. 110, New York,

Wiley 1978[3]; c) Salt, J. E., Girolami, G. S., Wilkinson, G., Motevalli, M., Thornton-Pett, M., Hursthouse, M. B.: J. Chem. Soc., Dalton Trans. 685 (1985); d) Carmona, E., Marin, J. M., Poveda, M. L., Atwood, J. L., Rogers, R. D.: J. ACS *105*, 3014 (1983); e) Uchida, T., Uchida, Y., Hidai, M., Kodama, T.: Acta Crystallogr., Sec. B *31*, 1197 (1975); f) Sato, M., Tatsumi, T., Kodama, T., Hidai, M., Uchida, T., Uchida, Y.: J. ACS *100*, 4447 (1978); g) Dadkhah, H., Dilworth, J. R., Fairman, K., Kan, C. T., Richards, R. L., Hughes, D. L.: J. Chem. Soc., Dalton Trans. 1523 (1985); h) Ibers, J. A., Davis, B. R.: Inorg. Chem. *10*, 578 (1971); i) Carvalho, M. F. N. N., Pombeiro, A. J. L., Schubert, U., Orama, O., Pickett, C. J., Richards, R. L.: J. Chem. Soc., Dalton Trans. 2079 (1985); j) Chiu, K. W., Wong, W.-K., Wilkinson, G., Galas, A. M. R., Hursthouse, M. B.: Polyhedron *1*, 37 (1982); k) Davis, B. R., Ibers, J. A.: Inorg. Chem. *9*, 2768 (1970); l) Chaudret, B., Devillers, J., Poilblanc, R.: J. Chem. Soc., Chem. Commun. 641 (1983); m) Fergusson, J. E., Love, J. L., Robinson, W. T.: Inorg. Chem. *11*, 1662 (1972); n) Davis, B. R., Payne, N. C., Ibers, J. A.: Inorg. Chem. *8*, 2719 (1969); o) Hoffman, P. R., Yoshida, T., Okano, T., Otsuka, S., Ibers, J. A.: Inorg. Chem. *15*, 2462 (1976); p) Thorn, D. L., Tulip, T. H., Ibers, J. A.: J. Chem. Soc., Dalton Trans. 2022 (1979); q) Sanner, R. D., Duggan, D. M., McKenzie, T. C., Marsh, R. E., Bercaw, J. E.: J. ACS *98*, 8358 (1976); r) Forder, R. A., Prout, K.: Acta Crystallogr., Ser. B *30*, 2778 (1974); s) Ziegler, M. L., Weidenhammer, K., Zeiner, H., Skell, P. S., Herrmann, W. A.: Angew. Chem. Int. Ed. Engl. *15*, 695 (1976); t) Mercer, M.: J. Chem. Soc., Dalton Trans. 1637 (1974); u) Cradwick, P. D.: J. Chem. Soc., Dalton Trans. 1934 (1976); v) Treitel, I. M., Flood, M. T., Marsh, R. E., Gray, H. B.: J. ACS *91*, 6512 (1969); w) Cecconi, F., Ghilardi, C. A., Midollini, S., Moneti, S., Orlandini, A., Bacci, M.: J. Chem. Soc., Chem. Commun. 731 (1985); x) Hammer, R., Klein, H.-F., Schubert, U., Frank, A., Huttner, G.: Angew. Chem. Int. Ed. Engl. *15*, 612 (1976); y) Yoshida, T., Okano, T., Thorn, D. L., Tulip, T. H., Otsuka, S., Ibers, J. A.: J. Organomet. Chem. *181*, 183 (1979)

electrons on either nitrogen atom ($:N \equiv N:$), and may be bonded to a metal atom as a unidentate ligand (end-on (i)), or link two metal atoms (end-on bridging (ii)). Since the discovery of the first dinitrogen complex $[Ru(NH_3)_5(N_2)]^+$ in 1965,[141] more than a hundred N_2 complexes have been prepared. Many of these exhibit bonding mode (i) while some show mode (ii). Table II lists some examples of these complexes whose structures have been determined by X-ray analysis.

3.3.1 End-on Unidentate Coordination

Dinitrogen is isoelectronic with CO and the bonding mechanism is also synergic, but its $(\sigma + \pi)$-bonding capacity seems to be weaker than that of CO. Thus, metal complexes of N_2 generally carry one or two terminal N_2 ligands, in contrast to the prevalence of polycarbonyl complexes. As is seen in Table II, the N—N distances observed in complexes of the end-on unidentate N_2 are not appreciably different from that of free dinitrogen (1.0976 Å).

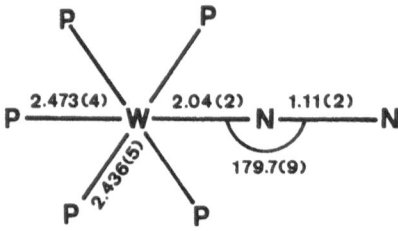

Fig. 24. Molecular structure of $W(N_2)(PMe_3)_5$.[142]

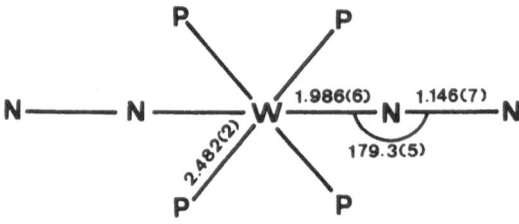

Fig. 25. The core of a molecule of trans-$W(N_2)_2(PEt_2Ph)_4$.[143]

Figures 24 and 25 show the molecular structures of recent examples of mono- and bis-dinitrogen complexes, respectively. The dispersed-sodium reduction of $WCl_4(PMe_3)_3$ in THF under N_2 gave cis-$W(N_2)_2(PMe_3)_4$, which upon reaction with PMe_3 under argon at 40–50 °C resulted in $W(N_2)(PMe_3)_5$.[142] Similarly, trans-$W(N_2)_2(PEt_2Ph)_4 \cdot (THF)$ was obtained together with other N_2-containing complexes from magnesium reduction of WCl_6 or $WCl_4(PEt_2Ph)_2$ in THF under N_2 in the presence of PEt_2Ph.[143]

The first tris(dinitrogen) complex isolated is mer-$Mo(N_2)_3\{PPh-(n-Pr)_2\}_3$ depicted in Fig. 26. This complex was prepared by the reaction

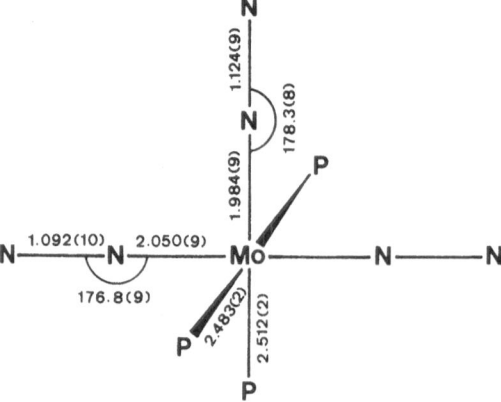

Fig. 26. Molecular geometry of mer-$Mo(N_2)_3\{PPh(n-Pr)_2\}_3$.[144]

of N_2 in benzene or THF with *trans*-$Mo(N_2)_2\{PPh(n\text{-}Pr)_2\}_4$ which had been produced by magnesium reduction of $MoCl_5$ or $MoCl_3(THF)_3$ in THF under N_2 in the presence of $PPh(n\text{-}Pr)_2$.[144]

3.3.2 End-on Bridging

The zirconium(II) complex $[(\eta\text{-}C_5Me_5)_2Zr(N_2)]_2(\mu\text{-}N_2)$ was prepared by reduction of $(\eta\text{-}C_5Me_5)_2ZrCl_2$ with excess sodium amalgam in toluene at room temperature under 1 atm of N_2. As is shown in Fig. 27, it has two terminal and one bridging N_2 ligands both bonded in the end-on fashion with essentially linear $ZrN{\equiv}N$ and $ZrN{\equiv}NZr$ arrangements.

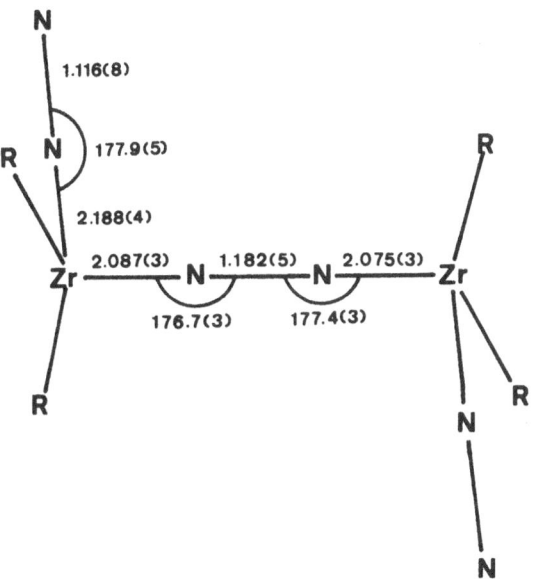

Fig. 27. Skeletal view of $[(\eta\text{-}C_5Me_5)_2Zr(N_2)]_2(\mu\text{-}N_2)$ with R representing an η-C_5Me_5 ring.[145]

This complex thus allows a comparison between the two coordination modes of N_2. The terminal N—N distances, 1.116(8) and 1.114(7) Å are longer than 1.0976(1) Å for gaseous N_2, and the bridging N—N distance of 1.182(5) Å is still longer, indicative of a substantial reduction in bond order. Of the three infrared absorption bands assignable to the N—N stretching modes which were recorded in KBr pellets, the lowest-frequency band (1578 cm^{-1}) of medium intensity may be attributed to the bridging N_2, while the strong bands at 2040 and 2003 cm^{-1} to the terminal N_2.[145]

The bridging N—N distances in

[W(PhC≡CPh)(dme)Cl$_2$]$_2$(μ-N$_2$)
(dme = 1,2-dimethoxyethane),[146]
[Ta(=CHCMe$_3$)(CH$_2$CMe$_3$)(PMe$_3$)$_2$]$_2$(μ-N$_2$),[147]
and
[TaCl$_3${P(CH$_2$Ph)$_3$}(THF)]$_2$(μ-N$_2$)[148]

are 1.292(16), 1.298(12), and 1.282(6) Å, respectively. These are much longer than usual N—N bond distances in other simple μ-N$_2$ complexes (Table II), and it was suggested that formally the (N—N)$^{4-}$ ion is involved to form the M=N—N=M linkage in these complexes. In fact the tantalum complexes react with HCl in ether to give N$_2$H$_4$(HCl) in high yields,[149] and [W(PhC≡CPh)(OCMe$_3$)$_2$]$_2$(μ-N$_2$) related to the above tungsten complex also reacts with HCl in ether to afford hydrazine in an essentially quantitative reaction.[146]

3.3.3 End-on μ_3-Bridging

Two complexes containing a $\mu_3(\eta^1, 2\eta^1)$-N$_2$ ligand (type (iii) in Fig. 23) have been reported to far: [WCl(py)(PMe$_2$Ph)$_3$(μ_3-N$_2$)AlCl$_2$]$_2$ · C$_6$H$_6$ and [(Me$_3$P)$_3$Co(μ_3-N$_2$)AlMe$_2$]$_2$, both of which contain Al(III) Lewis acids as the electron-pair acceptor.

The tungsten complex was prepared by treatment of cis-W(N$_2$)$_2$(PMe$_2$Ph)$_4$ with a twice molar amount of AlCl$_3$ followed by addition of 2 mol equivalent of pyridine in benzene.[150] The X-ray molecular structure is shown in Fig. 28. Each N$_2$ molecule end-on bridges one W and two Al atoms, and the N—N bond distance is as long as 1.46(4) and 1.25(3) Å. The IR bands observed at 1400 and 1350 cm^{-1} are assigned to the v(N$_2$) vibration. Two W, four N, and two Al atoms almost lie in the same plane. The Al—N bond is rather strong: the complex reacts with 1 mol equivalent of HCl gas in benzene to afford an aluminum hydrazido(2-) complex

[WCl(py)(NNHAlCl$_2$)(PMe$_2$Ph)$_3$]Cl

as crystals in moderate yield.[150]

The latter cobalt complex was prepared by the reaction between K[Co(N$_2$)(PMe$_3$)$_3$] and Me$_2$AlCl in ether at −78 °C.[151] The complex is very sensitive to oxygen and moisture but otherwise fairly unreactive. As is seen in Fig. 29, it has a dimeric structure quite similar to that of the above tungsten complex. The Co—N bond is short (1.642(4) Å) as

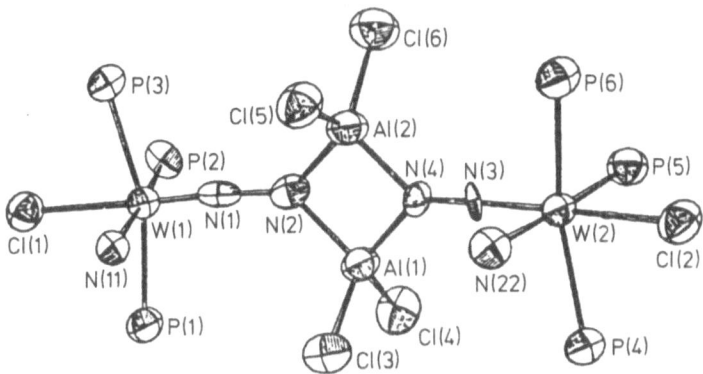

Fig. 28. Molecular structure of $[WCl(py)(PMe_2Ph)_3(\mu_3\text{-}N_2)AlCl_2]_2$. Important bond lengths (Å) and angles (°): N(1)—N(2) 1.46(4), N(3)—N(4) 1.25(3), W(1)—N(1) 1.64(4), W(2)—N(3) 1.82(4), W(1)—N(1)—N(2) 175(2), W(2)—N(3)—N(4) 178(2).[150]

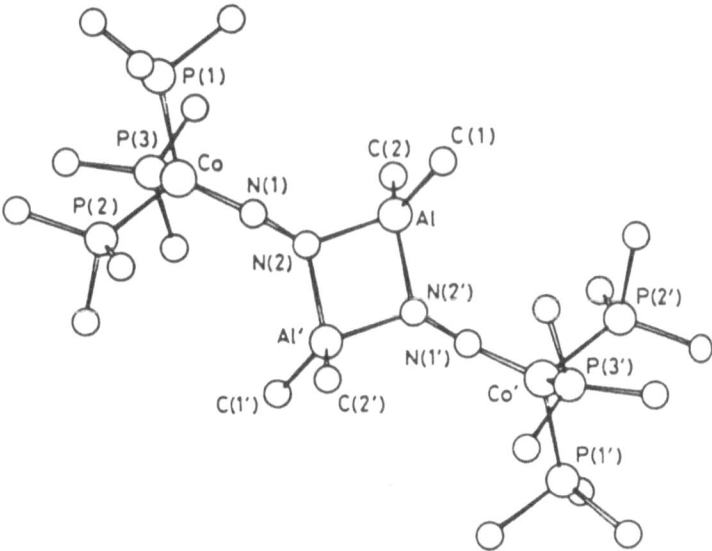

Fig. 29. Molecular structure of $[(Me_3P)_3Co(\mu_3\text{-}N_2)AlMe_2]_2$. The Co—N and N—N distances are 1.642(4) and 1.252(6) Å, respectively, and the Co—N—N angle is 171.5(4)°.[151]

compared with that in the parent mononuclear complex $[Co(N_2)(PMe_3)_3]^-$ (1.70 Å), and the N—N bond is long (1.252(6) Å). Thus it is claimed to be the first structurally characterized diazenido-bridged compound.[151]

3.3.4 Side-on Coordination

A paramagnetic complex $(\eta\text{-}C_5H_5)_2Zr(N_2)\{CH(SiMe_3)_2\}$ was prepared by sodium-amalgam reduction of $(\eta\text{-}C_5H_5)_2ZrCl\{CH(SiMe_3)_2\}$
in THF at 20 °C under N_2. It is the first N_2 complex of Zr(III) and was
claimed to be a type-(iv) complex (Fig. 23) based on the ESR-spectral
evidence for equivalence of two nitrogen atoms (a triplet in the $^{15}N_2$
complex and a quintuplet in the $^{14}N_2$ complex) together with the
absence of IR-active bands assignable to $\nu(N_2)$.[152] Unfortunately,
however, no X-ray evidence has been obtained.

3.3.5 Side-on Bridging

Two type-(v) complexes (Fig. 23) have been reported, both of which
contain nickel and lithium. An ethereal solution of phenyllithium and
Ni(CDT) (CDT = all-*trans*-1,5,9-cyclododecatriene) in the mole ratio of
3:1 absorbed N_2 (1 bar) at 0 °C to afford a dimeric complex *20* in ca. 50%
yield. The ether content in *20* varied in the range of 2-3 Et_2O per
6 PhLi:[153, 154]

$$4Ni(CDT) + 12PhLi + 2N_2$$
$$\rightarrow [\{(PhLi)_3Ni\}_2(N_2)(Et_2O)_{2-3}]_2 + 4CDT$$
$$20$$

When a mixture of PhLi and PhNa (1:3–4) was employed in place of
PhLi under similar reaction conditions,

$$[Ph(NaOEt_2)_2(Ph_2Ni)_2(N_2)NaLi_6(OEt)_4(Et_2O)]_2 \ (21)$$

was obtained among other products.[155]

Important structural subunits of complexes *20* and *21* are shown in
Figs. 30 and 31. In either case, an N_2 molecule side-on bridges two
diphenylnickel units, allowing a weak Ni—Ni bonding. Furthermore,
N_2 interacts in the end-on fashion with lithium atoms. In complex *20*, the
lone pair of one N atom interacts with just one Li ion, while that of the
other N atom is involved in a two-electron four-center bond with a Li_3
ring. In the case of *21*, similar two-electron three-center bonds to Li at
both ends of N_2 are observed.[156]

The N—N distances (1.35–1.36 Å) are greatly enlarged, lying within
the range of those observed for single and double bonds. On hydrolysis
of *20* in THF, some N_2 (30–38%) is converted into NH_3, and the rest
liberated as N_2 gas.[156]

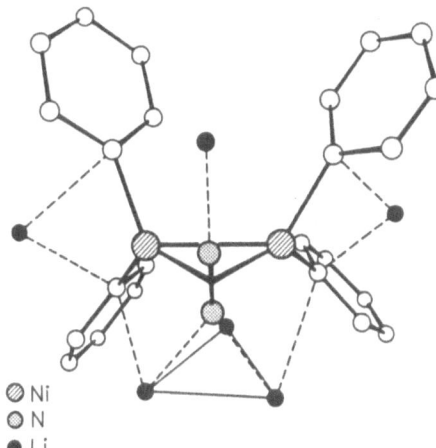

Ø Ni
◯ N
● Li

Fig. 30. Section of the molecule of [{(PhLi)₃Ni}₂(N₂)(Et₂O)₂]₂ *(20)* showing the interaction of dinitrogen with nickel and lithium.[156] Relevant bond distances (Å): N—N 1.35, Ni—N(av) 1.93, Li—N(av) 2.06, Ni—Ni 2.687.[154]

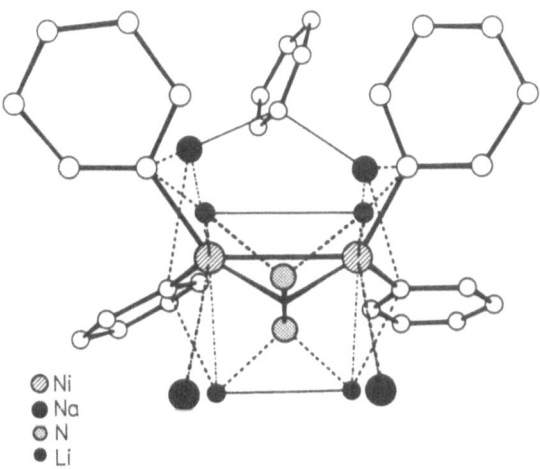

Ø Ni
● Na
◯ N
● Li

Fig. 31. Section of the molecule of [Ph(NaOEt₂)₂(Ph₂Ni)₂(N₂)NaLi₆(OEt)₄(Et₂O)]₂ *(21)* showing the interaction of dinitrogen with nickel and lithium.[156] Relevant bond distances (Å): N—N 1.359(18), Ni—N(av) 1.91(1), Li—N(av) 2.05(5), Ni—Ni 2.749(7).[155]

3.3.6 End-on:side-on μ_3-Bridging

A titanium complex containing a $\mu_3(2\eta^1, \eta^2)$-N$_2$ (type (vi) in Fig. 23) has been obtained by a series of reactions starting from $\mu(\eta^1:\eta^5\text{-}$ cyclopentadienyl)tris(η^5-cyclopentadienyl)dititanium(Ti-Ti), *22*:

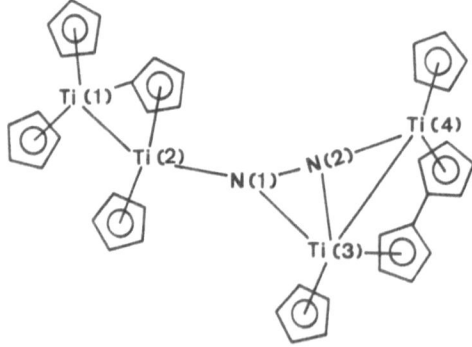

22

The red brown crystalline product has a $v(N_2)$ band at 1282 cm^{-1}, and its composition was determined by single-crystal X-ray analysis as $(\mu_3\text{-}N_2)$ [(η^5, η^5-C$_{10}$H$_8$) (η-C$_5$H$_5$)$_2$Ti$_2$] [(η^1, η^5-C$_5$H$_4$) (η-C$_5$H$_5$)$_3$Ti$_2$]-[(η-C$_5$H$_5$)$_2$(C$_6$H$_{14}$O$_3$)Ti] · C$_6$H$_{14}$O$_3$ (*23*) of which the essential skeleton is shown in Fig. 32. Here, C$_6$H$_{14}$O$_3$ is bis(2-methoxyethyl)ether (diglyme) which was used as a solvent.[157]

As a result of the multiple coordination, the N—N length (1.301(12) Å) is considerably long, being intermediate between that for azo (—N= N—) and hydrazo (>N—N<) compounds. Treatment of solutions of *23* in diglyme with dry HCl results in loss of most of the bound N$_2$. However, reaction of *23* with water in THF at 23 °C converts over 90% of the ligated N$_2$ to ammonia (mainly) and hydrazine.[157]

Fig. 32. Coordination skeleton of the $(\mu_3\text{-}N_2)$ [(η^5,η^5-C$_{10}$H$_8$) (η-C$_5$H$_5$)$_2$Ti$_2$]-[(η^1, η^5-C$_5$H$_4$) (η-C$_5$H$_5$)$_3$Ti$_2$] unit in *23*. Important bond lengths (Å) and angles (°): N—N 1.301(12), Ti(2)—N(1) 1.953(11), Ti(4)—N(2) 1.857(11), Ti(3)—N(1) 2.181(10), Ti(3)—N(2) 2.097(11), Ti(2)—N(1)—N(2) 145.6(9), Ti(4)—N(2)—N(1) 169.4(9).[157]

3.4 Protonation of the Coordinated Dinitrogen[139, 140]

Since the binding of N_2 to a transition metal is synergic, greater contribution from π backbonding than σ donation polarizes the ligand, increasing its basicity. Thus, a coordinated N_2 molecule, especially in Mo and W complexes, is protonated to afford ammonia and/or hydrazine, necessary electrons being supplied to N_2 from the central metal. The mechanisms of such protonation are not totally clear, but initial stages for the stepwise reduction are suggested in Scheme 4.

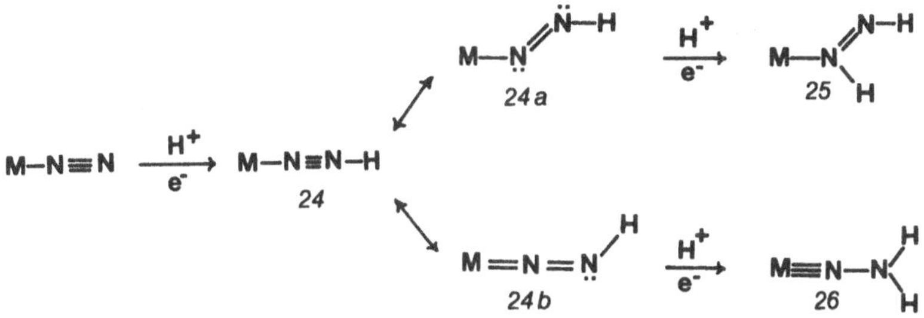

Scheme 4. Initial stages in the reduction of coordinated dinitrogen.[140]

Protonation on a unidentate N_2 results in the diazenido ligand N_2H *(24)*, which may be rearranged to *24a* or *24b* to generate a basic site at either N_α or N_β, respectively. Upon the second protonation, a diazene HNNH *(25)* or a hydrazido(2-)NNH$_2$ *(26)* ligand is produced.

For example, the reactions of *trans*-$M(N_2)_2(dppe)_2$ (M – Mo, W) with an excess amount of hydrogen halide in THF at room temperature followed by treatment with a salt of noncoordinating anion result in the hydrazido(2-) complexes:[158]

$$trans\text{-}M(N_2)_2(dppe)_2 \xrightarrow[-N_2]{2\,HX} MX_2(N_2H_2)(dppe)_2$$

$$\xrightarrow[-NaX]{NaY} [MX(NNH_2)(dppe)_2]Y$$

M = Mo; X = Br, Y = BF$_4$
M = W; X = Cl, Br, Y = ClO$_4$, PF$_6$, BPh$_4$

The X-ray structure of *trans*-[WCl(NNH$_2$)(dppe)$_2$] [BPh$_4$] indicates that the W—N—N linkage is essentially linear with the N—N distance (1.37(2) Å) corresponding to a bond order of 1.5 and the W—N distance (1.73(1) Å) near the triple bond length.[159]

The diazenido complexes, which should be produced by the reaction of the N_2 complex with 1 mol equivalent of HX, can not be isolated from the reaction mixtures. However, they are prepared cleanly by deprotonation of the hydrazido(2−) complexes in a medium in which the products are insoluble:[160]

$$trans\text{-}[WX(NNH_2)(dppe)_2]^+ \xrightarrow[\text{MeOH}]{\text{Et}_3\text{N}} trans\text{-}WX(N_2H)(dppe)_2$$

Treatment of the diazenido complexes with an equimolar amount of anhydrous acids regenerates the hydrazido(2-) compounds.[160]

Protonation of coordinated N_2 in complexes containing solely chelating diphosphines ends with formation of the hydrazido(2-) complexes, and further reduction is unattainable. On the contrary, N_2 complexes containing unidentate tertiary phosphines can give ammonia upon protonation. Thus, $cis\text{-}M(N_2)_2(PMe_2Ph)_4$ or $trans\text{-}M(N_2)_2(PMePh_2)_4$ (M = Mo, W) reacts with H_2SO_4 in MeOH at room temperature to liberate 1 mol N_2 gas and produce NH_3 in a high yield:

$$M(N_2)_2(PR_3)_4 + H_2SO_4 \xrightarrow[\text{MeOH}]{} N_2 + 2NH_3 + \text{other products}$$

$$(M = Mo, W; PR_3 = PMe_2Ph, PMePh_2)$$

The yield of NH_3 depends on the nature of the metal atom and is also sensitive to solvents. The Mo complexes give only about 0.65 mol NH_3 per mol complex, whereas the W complexes attain almost quantitative yields. Even on refluxing in MeOH without acids for 3−4 hours, $cis\text{-}W(N_2)(PMe_2Ph)_4$ yields about 1.6 mol NH_3 per mol complex.[161]

Protonation reactions of complexes containing the end-on bridging N_2 were investigated with $[(\eta\text{-}C_5Me_5)_2Zr(N_2)]_2(\mu\text{-}N_2)$ (Fig. 27). Treatment of this complex with a ten-times molar amount of dry HCl in toluene at $-80\,°C$ followed by warming to room temperature gave 0.86 mol of N_2H_4:[162]

$$[(\eta\text{-}C_5Me_5)_2Zr(N_2)]_2(\mu\text{-}N_2) + 4HCl \rightarrow 2(\eta\text{-}C_5Me_5)_2ZrCl_2 + 2N_2$$
$$+ N_2H_4(+ \text{small amount of } H_2)$$

Four electrons necessary for production of N_2H_4 from N_2 are supplied by two Zr(II) atoms. Since protonation of $[(\eta\text{-}C_5Me_5)_2Zr(^{15}N_2)]_2(\mu\text{-}^{14}N_2)$ under similar conditions gives an equimolar mixture of $^{15}N_2H_4$ and $^{14}N_2H_4$ almost quantitatively and

$[(\eta\text{-}C_5Me_5)_2Zr(CO)]_2(\mu\text{-}N_2)$ produces no trace of N_2H_4, it is evident that the unidentate N_2 ligand is indispensable for N_2H_4 production.[163]

The N_2 ligand bound to a metal atom may also be susceptible to attack by electrophilic reagents other than acids. In fact, reactions of coordinated N_2 or its protonated derivatives have been extensively studied with a range of organic compounds including alkyl and acyl halides, carboxylic anhydrides, aldehydes, ketones, and activated aryl or vinyl halides to directly form the carbon-nitrogen bonds.[164]

4 Triatomic Ligands

Among the triatomic ligands (incl. neutral molecules such as H_2O, SO_2, and CO_2) the anions: NO_2^-, N_3^-, NCO^-, NCS^-, $NCSe^-$, and $NCTe^-$ are well-known as "ambidentate" ligands. These anions have two or three donor atoms and can coordinate with a metal atom at one or the other position. Furthermore, they can serve not only as unidentate ligands but also as bridging ligands, connecting two or more metal atoms in various ways. This chapter focusses on the thiocyanate anion; the various coordination modes will be shown.

4.1 Coordination Modes for the Thiocyanate Ion

According to the MO calculation by Wagner,[165] the pi-electron structure of the thiocyanate ion may be represented as:

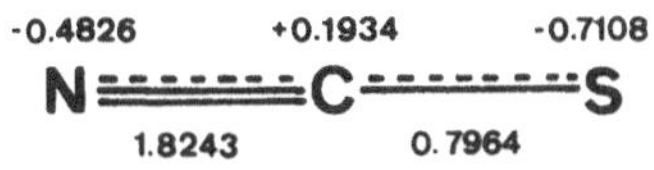

where the numbers written over the atoms are the resulting electron charges on the atoms, and the numbers written under the bonds are the pi-bond orders.

The difference in the charge densities on the N and S atoms is not very large. This may be the reason why the NCS^- anion exhibits a variety of coordination modes as shown in Fig. 33. There are a large number of metal complexes containing an N-bonded or S-bonded thiocyanate ion as a unidentate ligand. Various factors influencing the relative stabilities of these two bonding modes will be discussed in Section 4.3.

4.1.1 One-end Bridging

Two types of bridging modes, one-end bridging and end-to-end bridging, have been described so far. The following two subsections are

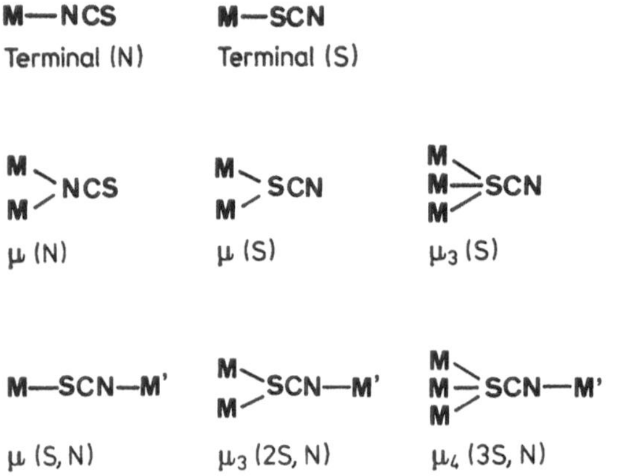

M—NCS   M—SCN

Terminal (N)   Terminal (S)

$M\diagdown NCS$ $M\diagdown SCN$ $M\diagdown SCN$
$M\diagup$ NCS $M\diagup$ SCN $M{=}SCN$ (M above)

μ (N)   μ (S) μ_3 (S)

M—SCN—M' $M\diagdown SCN{-}M'$ $M\diagup$ $M\diagdown SCN{-}M'$ $M{=}$

μ (S, N) μ_3 (2S, N) μ_4 (3S, N)

Fig. 33. Coordination modes for the NCS⁻ ion in metal complexes.

devoted to dinuclear complexes containing one-end μ(N)- and μ(S)-thiocyanate bridges.

4.1.1.1 μ(N) Bridging

Among the products of the reaction between sodium thiocyanate and $[n\text{-}Bu_4N]_3[Re_2Cl_8]$, a substance was found which was considered to involve CO ligands. X-Ray analysis later identified it to be $[n\text{-}Bu_4N]_3[Re_2(NCS)_{10}]$.[166]

The anion in this compound consists of two distorted octahedra sharing an edge formed by the N atoms of two bridging NCS⁻ ions (Fig. 34). The virtual D_{2h} symmetry of the inner Re_2N_{10} set of atoms implies that the unpaired electron in this formally mixed-valence Re(III)-Re(IV) complex is delocalized equally over both metal atoms. The Re—Re distance of 2.613(1) Å and the bridging angles of 102.7(3)° for N—Re—N and 77.4(3)° for Re—N—Re clearly indicate direct bonding, but the exact bond order is not certain.

Each NCS ligand, including the bridging ones, is effectively linear, the N—C—S angles ranging from 177.8 to 179.6°. For the eight terminal ligands, the Re—N—C angles range from 165.4 to 176.4°, and the average N—C and C—S distances are 1.16(1) and 1.59(1) Å. These data are typical for terminal NCS ligands, indicative of an electron distribution represented by the formal structure N≡C—S⁻. (The rather short C—S distance is explained by some pπ-dπ interaction.) The bridging NCS ligands are essentially perpendicular to the Re—Re' axis. The linearity and distances (N—C 1.17(1) Å, C—S 1.57(1) Å) in the bridge hardly allow to distinguish it from the terminal one. The IR absorptions at ca. 1900 cm⁻¹, which were the only data for proposing the formula

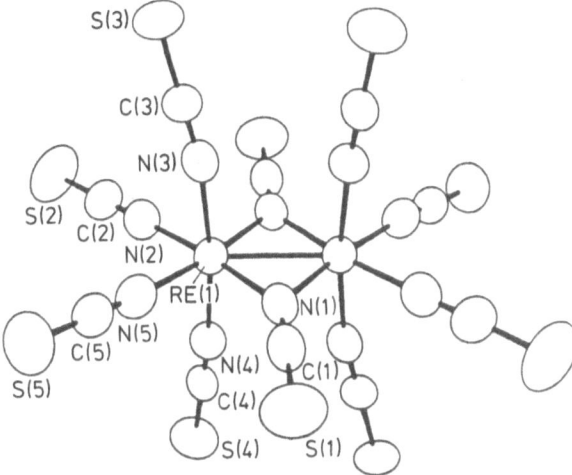

Fig. 34. A computer-generated drawing of $[Re_2(NCS)_{10}]^{3-}$ ion in the $[n\text{-}Bu_4N]^+$ salt.[166]

$[n\text{-}Bu_4N]_3[Re_2(NCS)_{10}(CO)_2]$, must be assigned to the CN stretching in the N-bridging NCS ligand.[166]

4.1.1.2 $\mu(S)$ Bridging

Figure 35 shows the molecular structure of the dicopper(I) complex $Cu_2L(SCN)_2$ of a 20-membered macrocyclic Schiff base ligand L which was prepared by the cyclic [2 + 2] condensation of two molecules of 2,5-

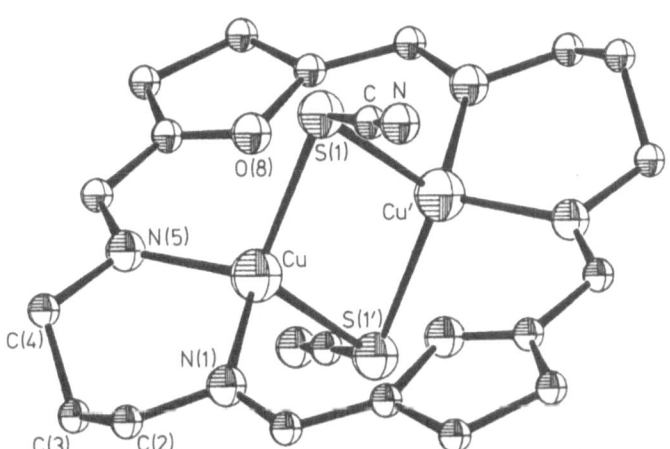

Fig. 35. Molecular structure of $Cu_2L(SCN)_2$, where L is a 20-membered 'N$_4$O$_2$' macrocyclic ligand.[167]

diformylfuran with two molecules of 1,3-diaminopropane. Each of the two tetrahedrally coordinated Cu(I) atoms is bonded to two nitrogen atoms of the macrocycle and to the two sulfur atoms of the bridging thiocyanate anions. The Cu_2S_2 plane is nearly perpendicular (85.4°) to the macrocycle plane. The N—C and C—S bond distances are 1.13(5) and 1.70(4) Å, respectively, and the Cu—S—C bond angles are 106.8° and 109.2°. These observations are consistent, as are the following IR data, with the predominance of the valence bond structure N≡C—S⁻: $v(CN)$ 2105 cm^{-1} (strong and sharp), $v(CS)$ 700 cm^{-1} (medium), and $\delta(NCS)$ 443, 462 cm^{-1} (weak).[167]

4.1.1.3 μ_3(S) Bridging

No discrete molecule containing covalent μ_3(S) linkage (Fig. 33) has been found as yet. The crystal structure of $NH_4Ag(SCN)_2$ is built up by AgSCN molecules, $NH_4{}^+$ ions, and NCS⁻ ions. The coordination around the S atom of a NCS⁻ ion is a distorted flat trigonal pyramid of three Ag and one C atoms. The Ag—S distances range within 2.630 to 2.742 Å and these bonds may be rather ionic.[168]

4.1.2 End-to-end Bridging

4.1.2.1 Single μ(S, N) Bridging

The dinuclear complexes $(NH_3)_5CoNCSCo(CN)_5$ and $(NH_3)_5{}^-$ $CoSCNCo(CN)_5$ involve a single bridging NCS group and constitute a couple of linkage isomers. The former more stable isomer was prepared by the reaction of equimolar amounts of $[(NH_3)_5CoNCS](ClO_4)_2$ and $K_2[Co(CN)_5(H_2O)]$ at 50°C for 3 hours, while the latter less stable isomer was obtained from the same reaction at 0°C. They were characterized by electronic and IR spectroscopy,[169] and the molecular strcture of the former complex was determined by X-ray analysis.[170] The bond distances (Å) and angles(°) related to the NCS group are as follows: Co—S 2,334(2), Co—N 1.908(4), N—C 1.145(7), C—S 1.645(5), Co—S—C 101.1(1), Co—N—C 169.5(2), and N—C—S 179.3 (2).

The crystal structure of $[HgSCN(\mu\text{-}SCN)\{P(C_6H_{11})_3\}]_\infty$ (Fig. 36) contains infinite chains of Hg atoms linked by singly bridging NCS groups. The Hg atom is situated 0.335 Å above the trigonal plane of S, S, and P. The bond length of Hg—S (bridging) (2.553(3) Å) is somewhat longer than Hg—S (terminal) (2.471(4) Å). An N atom from a neighboring molecule links to Hg (Hg—N 2.516(10) Å), constituting a distorted trigonal-pyramid configuration of Hg.[173]

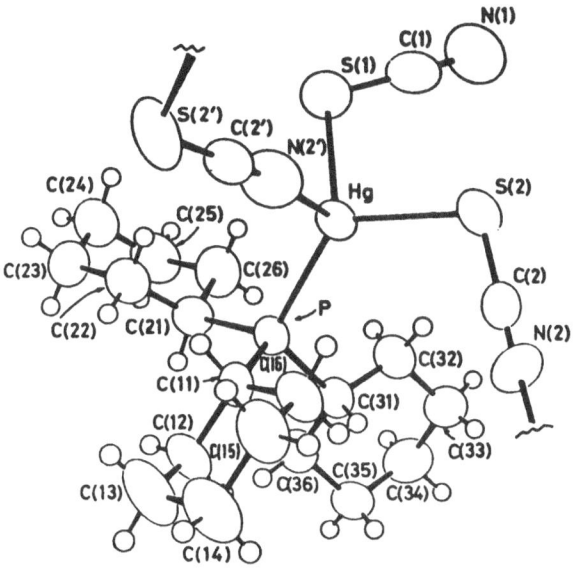

Fig. 36. Molecular structure of $[HgSCN(\mu\text{-SCN})\{P(C_6H_{11})_3\}]_\infty$.[173]

4.1.2.2 Double μ(S, N) Bridging

When $[PtCl(\mu\text{-Cl})\{P(n\text{-Pr})_3\}]_2$ is treated with a twice molar amount of KNCS in cold acetone solution, yellow $\alpha\text{-}[PtCl(\mu\text{-SCN})\{P(n\text{-Pr})_3\}]_2$ is obtained, while the same reaction carried out in the boiling solvent gives the pale greenish-yellow β-isomer.[171] They are both dimeric in boiling benzene, and were confirmed by X-ray analysis to be a couple of linkage isomers having the following structures:[172]

α-form   β-form

In either isomer, P, Cl, and the atoms of the eight-membered $\{Pt_2(SCN)_2\}$ ring are all approximately coplanar, two P atoms occupying the mutually *trans* positions. The bond lengths and bond angles related to the NCS group in these doubly bridged complexes (Table III) are quite similar to those for the singly bridged NCS cited above. The lengths of Pt—S and Pt—N bonds *trans* to $P(n\text{-Pr})_3$ are longer than those *trans* to Cl, reflecting the stronger *trans* influence of the tertiary phosphine than that of Cl⁻.

Table III. Bond Distances (Å) and Angles (°) Related to the NCS Group in the two Linkage Isomers of [PtCl(μ-SCN){P(n-Pr)$_3$}]$_2$[172]

	α-Isomer	β-Isomer
Pt—S	2.327(5)	2.408(4)
Pt—N	2.078(13)	1.965(13)
S—C	1.643(14)	1.641(15)
N—C	1.124(19)	1.168(16)
Pt—S—C	103.6(6)	102.9(5)
Pt—N—C	164.9(14)	167.3(11)
S—C—N	179.3(16)	178.6(13)

4.1.2.3 μ_3(2S, N) Bridging

In the case of Cd(SCN)$_2$, doubly bridging NCS groups connect Cd atoms to form infinite chains, which are packed so that sulfur atoms face Cd atoms of the nearest adjacent chains (Fig. 37). Thus, each Cd atom is

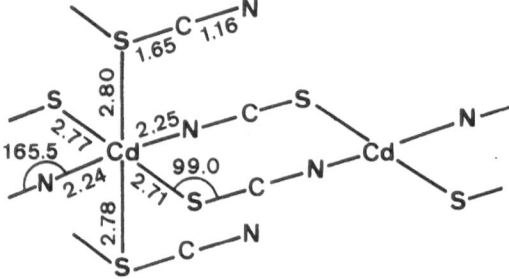

Fig. 37. Coordination skeleton around Cd and S in Cd(SCN)$_2$.[174]

octahedrally surrounded by four S and two *trans*-N atoms, while each NCS$^-$ ligand assumes the μ_3(2S, N) bridging mode (Fig. 33).[174]

4.1.2.4 μ_4(3S, N) Bridging

Copper(II) sulfate reacts with NH$_3$ and NH$_4$NCS in aqueous solution at room temperature to deposit crystals of Cu(NCS)$_2$(NH$_3$)$_2$. When the crystals are kept in the parent solution at 50°C for 2–3 weeks, Cu(II) is reduced to yield brown β modification of CuNCS, which is then transformed to the white α-CuNCS. In the crystal structure of α-CuNCS (Fig. 38), each Cu atom is tetrahedrally coordinated by 3S and N atoms,

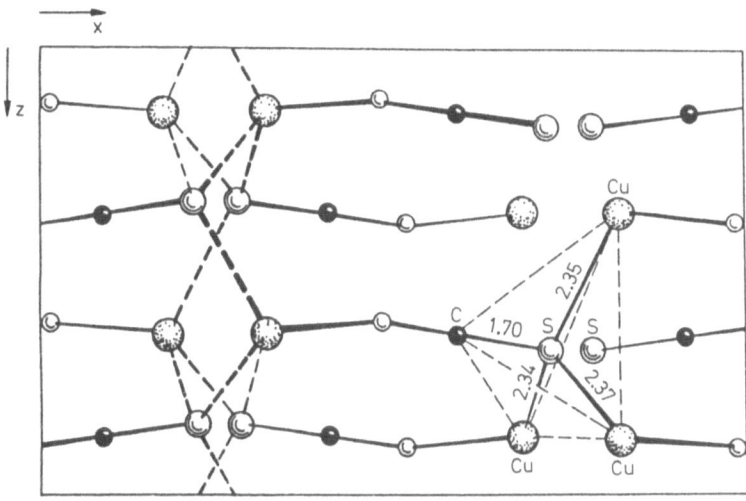

Fig. 38. Crystal structure of α-CuNCS in the (010) projection.[175]

the Cu—S distances being in the interval 2.34–2.37 Å and Cu—N being 1.93 Å. Each S atom is also tetrahedrally coordinated by 3 Cu and C atoms (S—C 1.70 Å).[175]

4.2 Infrared Spectroscopy for Determining the Coordination Modes of the Thiocyanate Ligand

Although X-ray crystallography is the most reliable technique for determining the coordination modes in crystals, various physical methods such as IR, Raman, electronic, NMR, NQR, and EPR spectroscopy as well as ESCA and mass spectrometry, have been employed with widely varying degrees of success.

Among these techniques, infrared spectroscopy has been most frequently used for determining the coordination modes of NCS^-.[176] The $v(CN)$, $v(CS)$, and $\delta(NCS)$ frequencies as well as integrated intensities of the $v(CN)$ band have been utilized. Table IV summarizes the frequency ranges for different types of NCS^- coordination.

The contributions of three resonating structures of NCS^- are estimated as: $N{\equiv}C{-}S^-$ 71%, $N{=}C{=}S$ 12%, and $^{-2}N{-}C{\equiv}S^+$ 17%.[177] Coordination through N increases the contribution of the third structure, increasing the C—S bond order and the $v(CS)$ frequency, whereas S-coordination increases the contribution of the first structure, increasing the C—N bond order and the $v(CN)$ frequency.

Table IV. Approximate Frequency Ranges and Integrated Intensity of the $v(CN)$ Band for Different Types of NCS^- Coordination[4]

Compound type	$v(CN)*$ cm^{-1}	$v(CS)*$ cm^{-1}	$\delta(NCS)$ cm^{-1}	$A \times 10^{-4}**$ M^{-1} cm^{-2}
NCS^-	2053	746	486, 471	3–5
M—NCS	2100–2050, s, b	870–820 w	485–475	7–11
M—SCN	2130–2085, s, sp	760–700 b	470–430	1–3
M—NCS—M	2165–2065	800–750	470–440	

* s: strong, w: weak, b: broad, sp: sharp
** Integrated intensity $A = (\pi/2\,cl)\,[\log(I_0/I)]\Delta v_{1/2}$, where c = concentration in mol dm^{-3} (M), l = cell thickness, I/I_0 = fraction of transmitted light, and $\Delta v_{1/2}$ = half-height width of the $v(CN)$ band.

The $v(CS)$ frequency distinguishes reliably between M—NCS and M—SCN, but in most mixed-ligand complexes the relevant frequency region is obscured by ligand absorptions. The $v(CN)$ frequency is less trustworthy, but the integrated intensity of the $v(CN)$ band differs significantly in the two types of compounds, being in the order M— $NCS > NCS^- > M$—SCN.

At present, integrated intensity measurements are made on spectra obtained from Nujol mulls or K Br disks containing both a complex and a reference substance such as 1,4-dicyanobenzene. The internal standard ratio (ISR) is calculated as the ratio of integrated absorption intensities of the $v(CN)$ band of the complex and the 2235-cm^{-1} band of the standard. ISR values of less than 10 are indicative of M—SCN coordination, while those characteristic of M—NCS coordination are larger than 20. Thus, the ISR values for trans-Pd(AsPh$_3$)$_2$(NCS)$_2$ and trans-Pd(AsPh$_3$)$_2$(SCN)$_2$ are 60.7 and 4.27,[178] and those for Fe(TIM)(NCS)$_2$ and [Fe(TIM)(SCN)$_2$]PF$_6$ are 21 and 3.2,[179] respectively. Here, TIM stands for a 14-membered tetraimine macrocycle, 2,3,9,10-tetramethyl-1,4,8,11-tetraazacyclotetradeca-1,3,8,10-tetraene, and S-bonding of both SCN ligands in [Fe(TIM)(SCN)$_2$]PF$_6$ was confirmed by X-ray crystallography. The observed S—C and C—N bond lengths are 1.677(5) and 1.148(7) Å, respectively, and the Fe—S—C angle is 104.0(2)°.[179]

4.3 Factors Influencing the Relative Stabilities of the N-bonded and S-bonded Thiocyanate Complexes[4, 5]

Since the coordination mode of NCS^- is intimately affected by the nature of the central metal ion and ancillary ligands as well as by the solvents and counterions, the NCS^- ligand has been used as a probe disclosing its chemical environment.

4.3.1 Principle of Hard and Soft Acids and Bases (HSAB)[180]

Pearson[180] noted that the relative strength of various bases is quite different depending upon the nature of the reference acid. The proton and CH_3Hg^+ were employed as two reference acids of different properties. Bases such as NH_3, en, OH^-, HPO_4^{2-}, and F^- in which the donor atom is N, O, or F prefer to coordinate with the proton. Bases such as PEt_3, S^{2-}, $S_2O_3^{2-}$, I^-, and CN^- in which the donor atom is P, S, I, or C prefer to coordinate with mercury.

The donor atoms in the former group are those which are of high electronegativity, of low polarizability, and difficult to oxidize. Bases containing these donor atoms are called "hard bases" to emphasize the fact that they hold on to their electrons tightly. The donor atoms involved in bases of the latter group are of low electronegativity, of high polarizability, and easy to oxidize. Bases containing these donor atoms are called "soft bases", a term which graphically describes the looseness with they hold their valence electrons. Table V shows a classification of bases into three categories: hard, soft, and borderline.[180]

An equivalent classification of Lewis acids into three categories was also proposed. When a Lewis acid is like the proton in preferring the hard bases and shows the following sequences of stabilities for com-

Table V. Classification of Lewis Bases[180]*

hard	soft	borderline
H_2O, OH^-, F^-	R_2S, RSH, RS^-	$C_6H_5NH_2$, C_5H_5N
$CH_3CO_2^-$, PO_4^{3-}, SO_4^{2-}	I^-, SCN^-, $S_2O_3^{2-}$	N_3^-, Br^-, NO_2^-
Cl^-, CO_3^{2-}, ClO_4^-, NO_3^-	R_3P, R_3As, $(RO)_3P$,	SO_3^{2-}, N_2
ROH, RO^-, R_2O, $S_2O_3^{2-}$	CN^-, RNC, CO, C_2H_4,	
NH_3, RNH_2, N_2H_4, NCS^-	C_6H_6, H^-, R^-	

* The *underlined* atom is the donor in cases where ambidentate coordination is possible.

plexes with different donor atoms, it is called a "hard acid":

$$N \gg P > As > Sb$$
$$O \gg S > Se > Te$$
$$F > Cl > Br > I$$

On the other hand, a Lewis acid which is like CH_3Hg^+ in preferring the soft bases and shows the following sequences of stabilities of complexes, is called a "soft acid":

$$N \ll P > As > Sb$$
$$O \ll S < Se \simeq Te$$
$$F < Cl < Br < I$$

Table VI gives the classification of a large number of Lewis acids into the three categories, hard, soft, and borderline.[180] Hard and soft acids correspond to and are extensions of "class a" and "class b" acids in the earlier criterion of Ahrland, Chatt, and Davies.[181] In general, acceptor atoms of hard acids are small in size, of high positive charge, and do not contain unshared electron-pairs in their valence shell (not all of these properties need be possessed by any one acid), leading to high electronegativity and low polarizability. On the contrary, soft acids have

Table VI. Classification of Lewis Acids[180]

hard	soft
H^+, Li^+, Na^+, K^+	Cu^+, Ag^+, Au^+, Tl^+, Hg^+
Be^{2+}, Mg^{2+}, Ca^{2+}, Sr^{2+}, Mn^{2+}	Pd^{2+}, Cd^{2+}, Pt^{2+}, Hg^{2+}, $MeHg^+$
Al^{3+}, Sc^{3+}, Ga^{3+}, In^{3+}, La^{3+}	$Co(CN)_5^{3-}$, Pt^{4+}, Te^{4+}, Tl^{3+}
N^{3+}, Cl^{3+}, Gd^{3+}, Lu^{3+}	$TlMe_3$, BH_3, $GaMe_3$, $GaCl_3$
Cr^{3+}, Co^{3+}, Fe^{3+}, As^{3+}, CH_3Sn^{3+}	GaI_3, $InCl_3$, RS^+, RSe^+, RTe^+
Si^{4+}, Ti^{4+}, Zr^{4+}, Th^{4+}, U^{4+}, Pu^{4+}	I^+, Br^+, HO^+, RO^+
Ce^{3+}, Hf^{4+}, WO^{4+}, Sn^{4+}	I_2, Br_2, ICN, etc.
UO_2^{2+}, Me_2Sn^{2+}, VO^{2+}, MoO^{3+}	trinitrobenzene, etc.
$BeMe_2$, BF_3, $B(OR)_3$, $AlMe_3$	chloranil, quinones, etc.
$AlCl_3$, AlH_3, RPO_2^+, $ROPO_2^+$	tetracyanoethylene, etc.
RSO_2^+, $ROSO_2^+$, SO_3	O, Cl, Br, I, N, $RO \cdot$, $RO_2 \cdot$
I^{7+}, I^{5+}, Cl^{7+}, Cr^{6+}, RCO^+, CO_2,	metal atoms, bulk metals
NC^+, HX (hydrogen bonding molecules)	

borderline
Fe^{2+}, Co^{2+}, Ni^{2+}, Cu^{2+}, Zn^{2+}, Pb^{2+}, Sn^{2+}, Sb^{3+}, Bi^{3+}
Rh^{3+}, Ir^{3+}, BMe_3, SO_2, NO^+, Ru^{2+}, Os^{2+}, R_3C^+, $C_6H_5^+$, GaH_3

acceptor atoms large in size, of low positive charge, and containing unshared pairs of electrons in their valence shell, leading to high polarizability and low electronegativity.

The Principle of Hard and Soft Acids and Bases (HSAB) proposed by Pearson relates the relative stabilities of acid-base complexes by stating that hard acids prefer to bind to hard bases and soft acids prefer to bind to soft bases.[180]

The most important factor which is relevant to the coordination mode of the thiocyanate ion is the nature of the central metal ion in accordance with HSAB. When NCS^- is coordinated with a hard metal (hard acid), it is bonded through N as a hard base, whereas it is bonded through S as a soft base to a soft metal (soft acid). Thus, the ligands in $[Co(NCS)_4]^{2-}$, $[Zn(NCS)_4]^{2-}$, and other binary thiocyanate complexes of the hard metals are exclusively N-bonded.

On the other hand, the NCS^- ions in $[M(SCN)_4]^{2-}$ and other binary complexes of soft metals such as Pd(II), Pt(II), and Hg(II) are all S-bonded. However, the coordination mode of NCS^- in mixed-ligand complexes is greatly affected by the nature of the ancillary ligands by virtue of their electronic and steric effects.

4.3.2 Electronic Effects of Ancillary Ligands

Characters of metals may be modified by the nature of the ligands bonded to the metal. For example, $[Co(NH_3)_5(NCS)]^{2+}$ is stable and $[Co(NH_3)_5(SCN)]^{2+}$ is much more unstable, whereas $[Co(CN)_5(SCN)]^{3-}$ is more stable than $[Co(CN)_5(NCS)]^{3-}$. These phenomena were regarded as a "symbiosis" by Jørgensen.[182] When hard ligands such as NH_3 are coordinated with a cobalt(III) ion, hardness of the central metal is not modified, preferring N-bonding of NCS^-. On the other hand, CN^- anions donate electrons more strongly, making Co(III) be so soft as to favor S-bonding of NCS^- over N-bonding. Thus, a flocking together of like ligands is observed.

Pearson noted that an "antisymbiotic effect" is also operative and explained this on the basis of a *trans* influence: two soft ligands in mutually *trans* positions will have a destabilizing effect on each other when attached to soft metal atoms.[183] Linearly dicoordinated LAu(CNS) complexes exemplify this effect. (Hereafter, CNS represents the thiocyanate ligand without specifying the coordination mode.) Although only the S-bonded isomer of each complex is present in the solid state, the N-bonded isomers of some complexes were found to coexist in a variety of solvents. The Au—NCS/Au—SCN ratio determined as the integrated intensity ratio of the $v(CN)$ absorption increases as the *trans* influence of L increases, being zero for L = ethylenethiourea, $S(CH_2Ph)_2$, and $AsPh_3$, and 0.52, 0,67, and 1.57 for L = PPh_3, PMe_3, and $P(OPh)_3$, respectively.[184]

In the case of the linear Au(I) complexes, the coordination mode of the thiocyanate ligand reflects solely the electronic effect of the *trans* ligand, being free from its steric effect. The stronger the *trans* influence of L, the harder the remaining site becomes and, consequently, the greater the Au—NCS/Au—SCN ratio.

Similarly,   Pd(1-diphenylphosphino-3-dimethylaminopropane)-(NCS)(SCN) *(27)* exhibits mixed bonding in crystals with the S-bonding *trans* to the amine group of the chelating ligand and the N-bonding *trans* to the phosphine group in accordance with the antisymbiotic effect.[185]

27

4.3.3 Steric Effects of Ancillary Ligands

When a thiocyanate anion is N-bonded to a metal atom, the M—NCS arrangement is essentially linear, whereas the M—S—C angle is about 100° for the M—SCN bonding. Thus, the S-bonding has a higher steric demand than the N-bonding, and is disfavored by adjacent bulky ligands.

For example, a series of Pd(II) complexes of Pd(diamine)(CNS)$_2$ containing bidentate substituted diamines representing a wide range of steric requirements were characterized by various spectroscopic measurements. In the solid state, the diamines producing least steric hindrance (1,2-diamino-2-methylpropane, 1,2-bis(methylamino)ethane, 1-amino-2-dimethylaminoethane, and 2-aminomethylpyridine) were found to form salts as [Pd(diamine)$_2$](SCN)$_2$. Diamines having intermediate steric demands   (1,2-bis-(dimethylamino)ethane,   1,2-bis(phenylamino)-ethane, 1-amino-2-morpholinoethane, 1-amino-2-pyrrolidinoethane, and 1-amino-2-piperidinoethane)   produced   S-bonded   complexes   of Pd(diamine)(SCN)$_2$, whereas those exhibiting the greatest steric requirements (1,2-dipiperidinoethane, 1,2-dipyrrolidinoethane, and 1,3-bis(dimethylamino)propane)   yielded   N-bonded   complexes   of Pd(diamine)(NCS)$_2$.[186]

The molecular structures of a series of complexes $\{Ph_2P(CH_2)_nPPh_2\}Pd(CNS)_2$ determined by X-ray diffraction techniques also reveal the steric effect on the coordination mode of NCS^-. As is illustrated in Fig. 39, the P—Pd—P angle and hence the steric effect increase with n, causing the change in the coordination mode of the NCS^- ligand.[187]

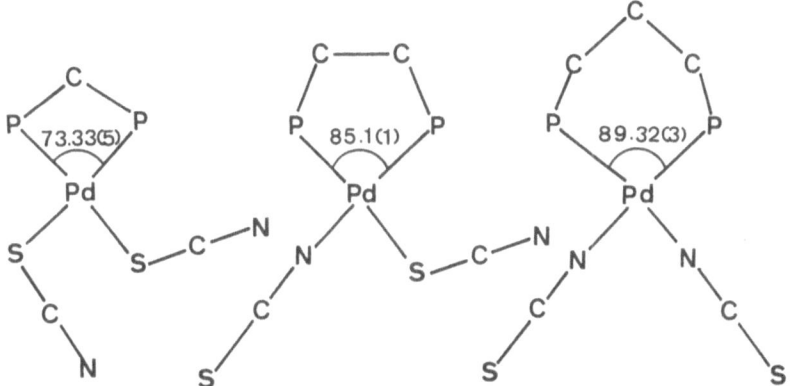

Fig. 39. Molecular skeletons of the $\{Ph_2P(CH_2)_nPPh_2\}Pd(CNS)_2$ complexes $(n = 1, 2,$ and $3).$[187]

In the case of the square-planar complexes $(R_3P)_2Pd(CNS)_2$, both the electronic and steric effects are important in determining both the coordination geometry and thiocyanate bonding mode. Electronic effects favor the *cis* configuration about Pd and the S-bonding mode unless steric effects are large. Steric effects, however, are the discriminator for both the geometry and CNS bonding mode. When steric effects become larger, the favored geometry becomes *trans* rather than the electronically-favored *cis* and the CNS bonding mode becomes N rather than the electronically favored S-bonding.

Examples are given by a series of complexes of $L_2Pd(CNS)_2$, where $L = 1$-R-3,4-dimethylphosphole, and $R = Me, t\text{-}Bu, Ph,$ or CH_2Ph.

These complexes have the *cis* geometry in the solid state with the exception of the bulkiest substituent $R = t\text{-}Bu$, which is *trans*. X-Ray analysis confirmed the S-bonding of both CNS ligands in *cis*-Pd(SCN)$_2$(1,3,4-trimethylphosphole)$_2$.[188] In the case of R = Ph, on the other hand, the *cis* geometry is preserved, but one of the CNS ligands is changed to N-bonding (Fig. 40).[189]

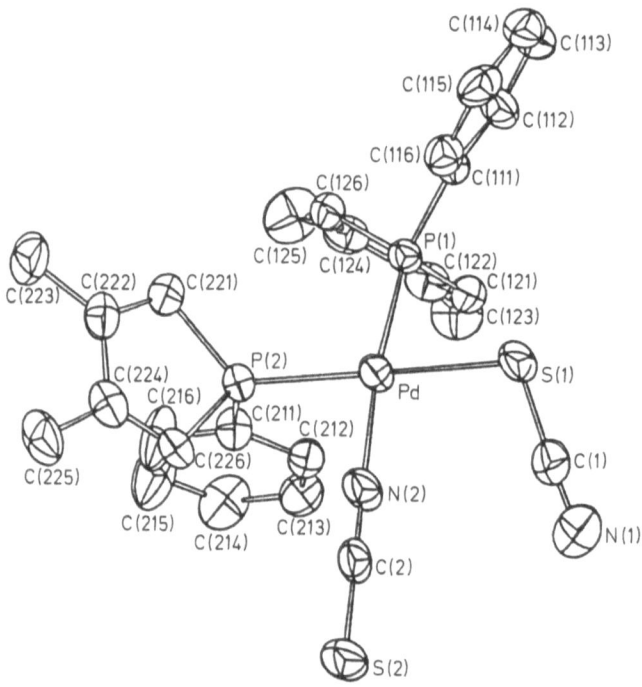

Fig. 40. Molecular structure of Pd(NCS)(SCN)(1-Ph-3,4-Me$_2$-phosphole)$_2$. Bond lengths (Å) and angles (°): Pd—S 2.385(2), S(1)—C(1) 1.651(9), C(1)—N(1) 1.137(13), Pd—N(2) 2.030 (5), N(2)—C(2) 1.151(7), C(2)—S(2) 1.619(6), Pd—S(1)—C(1) 102.8(3), S(1)—C(1)—N(1) 176.3(10), Pd—N(2)—C(2) 171.0(7), N(2)—C(2)—S(2) 178.8(8).[189]

Thus, the bonding mode of CNS seems to be more sensitive to small changes in steric effects than is the coordination geometry. Consequently, steric effects smaller than those necessary to stabilize *trans* geometry promote N-bonding.

4.3.4 Solvent and Counterion Effects

Relative stabilities of the thiocyanate linkage isomers in solution depend upon the nature of the solvent. Thus, square-planar Pd(II) and Pt(II) complexes of type ML$_2$(CNS)$_2$ (L = PPh$_3$, AsPh$_3$, SbPh$_3$, and other neutral ligands) adopt only the S-bonded mode in solvents such as DMF and DMSO which have relatively high dielectric constants. In solvents which in general have relatively low dielectric constants, e.g. C_6H_6 and $CHCl_3$, these complexes (with the exception of $Pd(phen)(SCN)_2$) exhibit either a mixture of S- and N-bonding modes or N-bonding alone (M—SCN—M bridging is also observed when $L=AsPh_3$ and $SbPh_3$).[190]

In aqueous solutions, $[Co(CN)_5(CNS)]^{3-}$ exists as an equilibrium mixture of the S-bonded (70%) and the N-bonded (30%) isomers. However, in CH_2Cl_2, $PhNO_2$, Me_2CO, and furfural, the N-bonded isomer is more stable.[191]

Electron paramagnetic resonance spectra recorded in frozen solutions of $[Fe(TIM)(SCN)_2]PF_6$ at 77 K show that the complex exists as a mixture of linkage isomers, although both SCN ligands are S-bonded in crystals (p. 66). As is shown in Table VII, the isomer ratio observed in neat acetone is close to the expected statistical distribution (S,S:S,N:N,N = 1:2:1). Increase in the water:acetone ratio dramatically decreases the amounts of N,N and N,S isomers relative to the S,S isomer.[179]

Table VII. Relative Amounts of S,S; S,N; and N,N Isomers of $[Fe(TIM)(SCN)_2]PF_6$ in Frozen Solution*[179]

Solvent, %		H_2O, %	S,S:S,N:N,N
MeNO$_2$	100	0	1.3:1.0:1.1
Me$_2$CO	100	0	1.2:2.0:0.8
	90	10	1.1:1.0:1.0
	76	24	2.3:1.1:1.0
	51	49	3.1:1.0

* TIM = 2,3,9,10-tetramethyl-1,4,8,11-tetra-azacyclotetradeca-1,3,8,10-tetraene (see text p. 66)

These observed results seem to accord with the supposition that a stronger hydrogen bond forms with M—SCN than with M—NCS. The nitrogen end of the NCS^- would be solvated to a greater extent via hydrogen bonding interaction since N is a first-row element. In agreement with this argument, thiocyanic acid exhibits a structure wherein the proton is bonded to the nitrogen.[191]

Several examples have been reported showing that the bonding mode in the solid state is affected by the nature of the counterion. Thus, $K_3[Co(CN)_5(NCS)]$ in KBr pellets isomerizes at 128 °C irreversibly to

$K_3[Co(CN)_5(SCN)]$, whereas $[(n\text{-}Bu)_4N]_3[Co(CN)_5(SCN)]$ completely isomerizes to the N-bonded isomer at room temperature within 3 days. The stabilization of the N-bonded isomer in this case was suggested to be due to an electronic effect in which the polarizable end of Co—NCS is better accommodated by the nonpolar, hydrocarbon environment of the $[(n\text{-}Bu)_4N]^+$ counterion.[192] The $[Me_4N]^+$ ion also stabilizes the N-bonded isomer, while the S-bonded isomer is favored by $[MeNH_3]^+$, Cs^+, and $[Co(NH_3)_6]^{3+}$ ions as well as K^+.[191]

An octahedral complex $trans\text{-}[IrH(piperidine)_4(NCS)]NCS$ involves an N-bonded NCS^- ligand. Although Ir(III) is a soft metal, the antisymbiotic effect exerted by the hydride ligand seems to disfavor the S-bonding. However, the bonding preference suffers counterion control. $trans\text{-}[IrH(piperidine)_4(SCN)][BPh_4]$ is more stable than the N-bonded isomer in the solid state, although it reverts to N-bonding on dissolution.[193]

5 Polyatomic Ligands: β-Dicarbonyl Compounds[194]

2,4-Pentanedione (acetylacetone) and other β-dicarbonyl compounds constitute a class of the most important polyatomic ligands, which have been employed very widely from the outset of this century.[195, 196] They are very versatile and exhibit a great variety of coordination modes besides the usual bidentate behavior of monoanions. This chapter briefly summarizes the coordination modes established so far for β-dicarbonyl compounds as the molecule, monoanion, and dianion (Fig. 41), with emphasis placed on the interconversions among them. A dinuclear Pt(II) complex containing an acetylacetonate trianion as a bridging ligand, which is now under detailed investigation, will also be described.

5.1 Three Coordination Modes for Neutral Molecules

5.1.1 Keto-enol Tautomerism and Structures of Enol Molecules

The β-dicarbonyl compound (β-dikH) generally exists as an equilibrium mixture of the tautomeric keto and enol forms. The rate of spontaneous interconversion between these forms is rather slow at room temperature,[197, 198] and their simultaneous NMR spectroscopic observation is possible. For instance, the ^{1}H NMR spectrum of neat acetylacetone (acacH) is composed of OH, CH, CH_2, and CH_3 signals in accordance with the following equilibrium (δ in ppm from internal Me_4Si):

$$\tag{8}$$

The equilibrium quotient $K = [\text{enol}]/[\text{keto}]$ for neat acacH is 3.59 (78.2% enol) at 37.3 °C and decreases with rising temperature, ΔH and

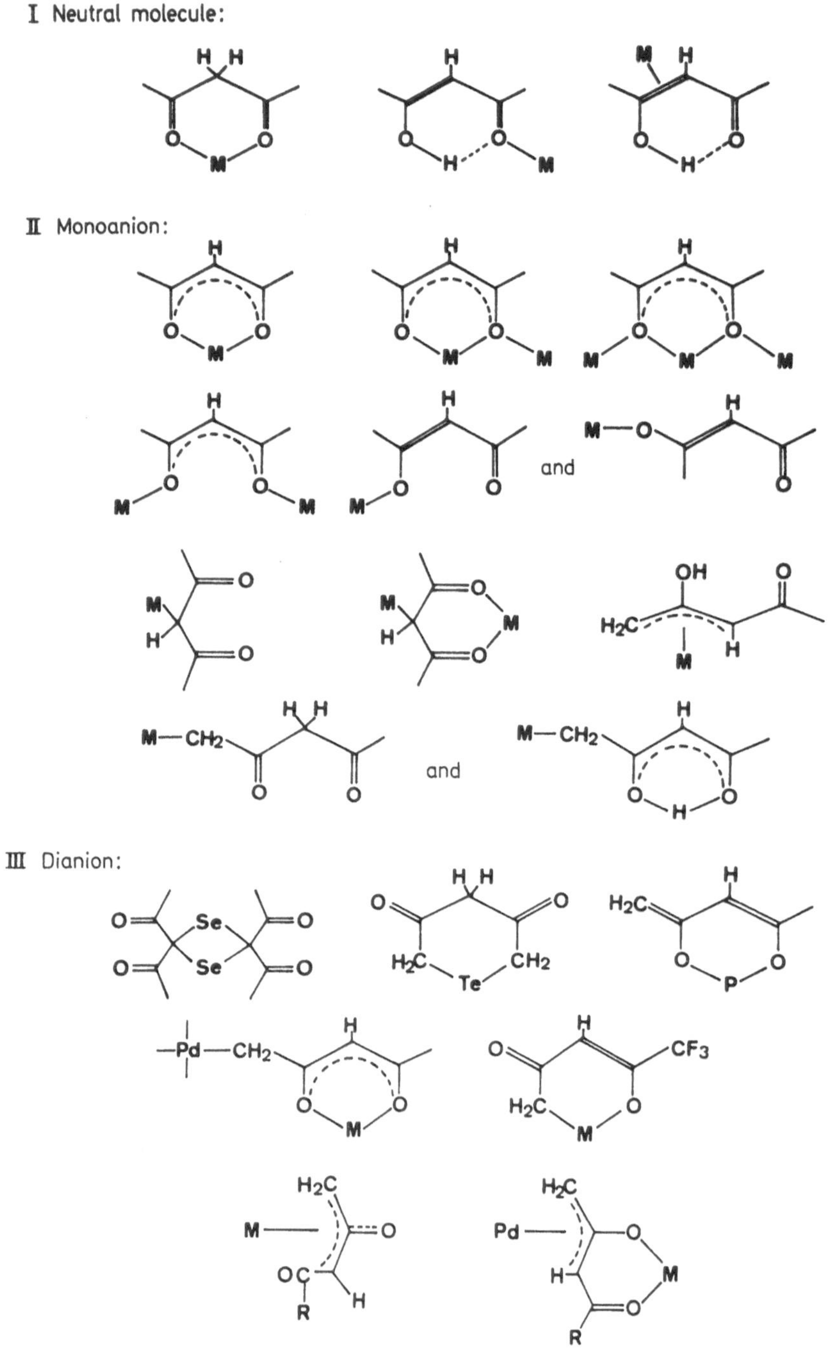

I Neutral molecule:

II Monoanion:

III Dianion:

Fig. 41. Coordination modes for β-dicarbonyl compounds as the molecule, monoanion, and dianion in metal complexes.

ΔS being -2.4 ± 0.2 kcal mol^{-1} and -5.2 ± 0.6 cal deg^{-1} mol^{-1}, respectively.[199] Thus, the enol form is more stable than the keto form owing to the intramolecular hydrogen bond.

Replacement of the terminal methyl groups of acacH by an electron-withdrawing or aromatic group such as CF_3, C_4H_3S (2-thienyl), and C_6H_5 shifts the equilibrium in favor of the enol tautomer. Hexafluoro-acetylacetone (hfacH), thenoyltrifluoroacetone (in carbon disulfide), and dibenzoylmethane (dbmH) (in carbon tetrachloride), for example, exist entirely in the enol form.[200-202]

Similarly, the electron-withdrawing substituent at the central atom also increases the enol content: 3-chloro- and 3-ethoxycarbonyl-2,4-pentanedione are composed of 84 and 100% enol, respectively. On the other hand, the electron-releasing substituents decrease the enol content. Thus, 3-methyl-, 3-ethyl-, and 3-isopropyl-2,4-pentanedione contain 29, 28, and 0% enol at 25°C, respectively.[203]

As is shown in Eq. (8), acacH exhibits only two methyl-proton signals at $\delta 1.97$ and $\delta 2.14$ assignable to the enol and keto methyls, respectively, indicating that the two methyl groups in an enol molecule are equivalent. The enol tautomer must either have the symmetric C_{2v} structure with a single energy minimum or exist in the asymmetric C_s forms with a low barrier to interconversion via the C_{2v} structure:

$$C_s \qquad\rightleftharpoons\qquad C_{2v} \qquad\rightleftharpoons\qquad C_s$$

Acetylacetone is a liquid at room temperature (m.p. $-23.2°C$, b.p. $140.5°C$) and has not been subjected to crystallographic examination. Camerman et al.[204] crystallized a complex of diphenylhydantoin and 9-ethyladenine (antiepileptic drug) from acacH solution and found that one enol molecule of acacH per asymmetric unit is involved in the crystal lattice. X-Ray analysis disclosed the structure of the enol molecule in a rather isolated environment, since no contacts shorter than van der Waals distances were observed between acacH atoms and those of diphenylhydantoin and ethyladenine.

As is seen in Fig. 42a, bond lengths, including the hydrogen bond are not symmetric, but are indicative of localized single and double bonds throughout the molecule. The $O \cdots O'$ distance is 2.535 Å.

Very recently, Shibata and coworkers[205] revealed by electron diffraction that the molecular structure of the enol form of gaseous acacH is not

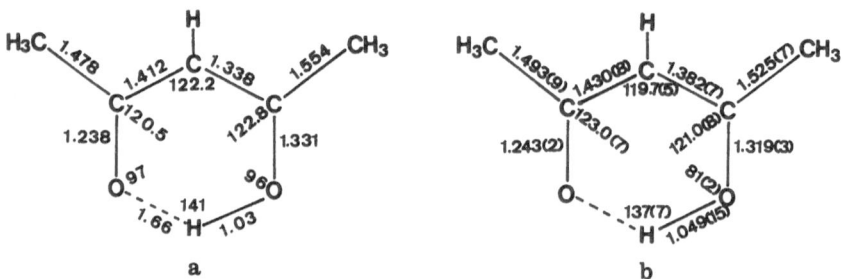

Fig. 42 a, b. Bond lengths (Å) and angles (°) for acetylacetone in a crystal (a) determined by X-ray (with esd's of 0.001 Å and 0.9°)[204] and in the gas phase (b) determined by electron diffraction.[205]

symmetric, either. The molecular skeleton as depicted in Fig. 42 b is quite similar to that in a crystal (Fig. 42 a). The O ··· O' distance is 2.512(8) Å, being slightly shorter than that in the solid phase owing to the stronger hydrogen bond in the gas phase.

The O_{1s} ionization regions of malonaldehyde, hfacH, and other related compounds were studied by X-ray photoelectron spectroscopy. Since these determinations are made on isolated gas-phase molecules on a time scale of 10^{-16} s, they may provide experimental evidence for the molecular symmetry. All of these compounds show two dominant ionizations arising from oxygens, indicating that the two oxygen atoms are made inequivalent by virtue of an asymmetric hydrogen bond.[206]

The crystal structure of dbmH has been determined by both X-ray[207] and neutron[208] diffraction methods. The intramolecular hydrogen bond is asymmetric, the difference in the O—H bond length being 0.199(17) Å.[208] The O ··· O' distance is 2.459(4) Å. Similarly, tetraacetyl-ethane,[209] 3,3'-dithiobis(2,4-pentanedione),[210] 3,3'-trithiobis(2,4-penta-nedione),[211] 2,2'-dithiobis(1-phenyl-1,3-butanedione),[212] and thenoyl-trifluoroacetone[213] have also been reported to exist in the enol form which contains an asymmetric hydrogen bond.

A neutron diffraction study on benzoylacetone (bzacH)[214] and X-ray studies on p-bromobenzoylacetone[215] and p-nitrobenzoylacetone[216] performed by Jones indicate that either molecule exists in the enol form and the intramolecular hydrogen bond is short with O ··· O' being 2.489(5), 2.481(9), and 2.457(5) Å, respectively. However, reliable evidence for asymmetry of the internal hydrogen bond was not obtained in these cases because of large thermal motion.

5.1.2 Metal Complexes of Neutral Molecules

Various metal complexes containing β-dikH as a neutral ligand have been isolated (Table VIII), although they are not, in general, very stable. They were prepared by three methods:

(a) Addition of a β-dikH molecule to a simple metal salt or a coordinatively unsaturated metal complex:

$$MX_n + \beta\text{-dikH} \rightarrow MX_n(\beta\text{-dikH})$$

(b) Substitution of a weak neutral ligand with β-dikH:

$$ML + \beta\text{-dikH} \rightarrow M(\beta\text{-dikH}) + L$$

(c) Addition of a proton to an anionic β-dik ligand:

$$M(\beta\text{-dik}) + HX \rightarrow MX(\beta\text{-dikH}) \text{ or } M(\beta\text{-dikH})^+ + X^-$$

Table VIII. Isolated Metal Complexes Containing a β-Dicarbonyl Compound as a Neutral Ligand[194]

Complex	Method of prepn[1]	Coord. mode of β-dikH[2]	ν(CO) for β-dikH in Nujol
$[Mg(acacH)_2(H_2O)_2](ClO_4)_2$	(a)	(1)	
$TiCl_4(acacH)$	(a)	(1)	1683, 1633
$TiX_4(3,3\text{-Me}_2\text{-acacH})$	(a)	(1)	1680–1691,
$(X = Cl, Br, I)$			1614–1621
$ZrCl_4(3,3\text{-Me}_2\text{-acacH})$	(a)	(1)	1644, 1597
$SnCl_4(acacH)$	(a)	(1)	1731, 1670
$SnCl_4(3\text{-Me-acacH})$	(a)	(1)	1673, 1613
$SnCl_4(3,3\text{-Me}_2\text{-acacH})$	(a)	(1)	1680, 1607
$[VO(ACOAP)(acacH)] \cdot H_2O^3$	(a)	(1)	1700
$[VO(ACAA)(acacH)(H_2O)]$			
$\cdot H_2O^4$	(a)	(1)	1700
$CrCl_2(acac)(acacH)$	(b)	(1)	1739 sh, 1692
$CrBr_2(acac)(acacH)$	(c)	(1)	1680
$MoOCl_3(acacH)$	(a)	(1)	1695, 1634
$MoO_2Cl_2(acacH)$	(a)	(1)	1695
$MoO_2Cl_2(etmalH)$	(a)	(1)	1695
$MnBr_2(acacH)_2$	(a)	(2)	1627
$MnBr_2(etmalH)$	(a)	(1)+(2)	1753 sh, 1726, 1658
$[ReCl(acacH)(CO)_3]_2$	(b)	(2)	
$[ReCl(bzacH)(CO)_3]_2$	(b)	(2)	1590
$CoCl_2(acacH)$	(a)	(1)	1727 sh, 1703
$CoCl_2(etacH)$	(a)	(1)+(2)	1717 sh, 1702, 1657
$CoCl_2(etmalH)$	(a)	(1)+(2)	1750 sh, 1729, 1698, 1657

Continuation next page

Table VIII (Continuation)

Complex	Method of prepn[1]	Coord. mode of β-dikH[2]	$\nu(CO)$ for β-dikH in Nujol
CoBr$_2$(acacH)	(a), (c)	(1)	1720sh, 1705
CoBr$_2$(etacH)	(a)	(1)+(2)	1720sh, 1703, 1685, 1650
CoBr$_2$(etmalH)	(a)	(1)+(2)	1755sh, 1726, 1690, 1655
[Ni(acacH)$_3$](ClO$_4$)$_2$	(b)	(1)	1700
[Ni(acacH)$_2$(H$_2$O)$_2$](ClO$_4$)$_2$	(b), (c)	(1)	1700
[Ni(acacH)$_2$(MeCO$_2$H)$_2$] (ClO$_4$)$_2$	(b)	(1)	1700
NiBr$_2$(acacH)$_2$	(c)	(1)	1729sh, 1693
PtCl(acac)(acacH)	(c)	(3)	1627
Pt(PPh$_3$)$_2$(acacH)	(a)	(3)	1682
Pt(PPh$_3$)$_2$(etmalH)	(a)	(3)	1681
ZnCl$_2$(acacH)	(a)	(1)	1715
[Zn(acacH)$_2$(H$_2$O)](ClO$_4$)$_2$	(b)	(1)+(2)	1725, 1700, 1620
ZnCl$_2$(etacH)	(a)	(1)+(2)	1730, 1706, 1625
ZnCl$_2$(etmalH)	(a)	(1)+(2)	1737, 1715, 1683sh, 1618
HgCl$_2$(acacH)	(a)	(1)	1698, 1658
Eu(tta)$_3$(acacH)[5]	(a)	(1)	1715
UO$_2$(acac)$_2$(acacH)	(a)	(2)	ca. 1620

[1] See text p. 79
[2] See text p. 80
[3] ACOAP-H$_2$ = CH$_3$C(OH)=CHC(CH$_3$)=NC$_6$H$_4$OH(o)
[4] ACAA-H$_2$ = CH$_3$C(OH)=CHC(CH$_3$)=NC$_6$H$_4$CO$_2$H(o)
[5] tta = 1-(2-thienyl)-4,4,4-trifluoro-1,3-butanedione (thenoyltrifluoroacetone)

In some cases, reduction of the central metal ion is accompanied:[217, 218]

$$2\,MoOCl_3 + 3\,acacH \rightarrow 2\,MoOCl_2(acacH) + HCl + acacCl$$
$$Co(acac)_3 + 2\,HBr + RH \rightarrow CoBr_2(acacH) + 2\,acacH + R\cdot$$

There are three coordination modes proposed for β-dikH: (1) O,O'-chelation of a keto tautomer, (2) O-unidentate linkage of an enol form, and (3) η^2(C,C') coordination of an enol molecule.

The tautomeric form of β-dikH is most conveniently diagnosed by infrared spectroscopy. The IR absorption bands in the regions of carbonyl and metal-oxygen stretching vibrations are especially helpful.

Thus, free acacH exhibits three carbonyl bands at 1730, 1712, and 1630 cm^{-1}. The higher frequency bands are attributed to the keto tautomer and the lowest-frequency band to enol.

5.1.3 O,O'-Chelation of the Keto Molecules

As is seen in Table VIII, the majority of complexes of β-dikH show the $v(C=O)$ band at around 1700 cm^{-1} and are supposed to contain a ketonic acacH molecule as a chelating ligand. Only two compounds among them have been confirmed so far by X-ray analysis to have the proposed structure. A light green crystal of NiBr$_2$(acacH)$_2$ is composed of discrete octahedral molecules of the *trans* configuration, acacH molecules forming an O,O'-chelate ring of boat conformation.[219]

The crystal structure of [Ni(acacH)$_2$(H$_2$O)$_2$](ClO$_4$)$_2$ was first reported by Anzenhofer and Hewitt[220] and reinvestigated later by Cramer et al.[221] There are two nonequivalent, centrosymmetric cations per unit cell. In both cations the acacH ligand is coordinated in the keto form, but differs in the conformation of the six-membered chelate ring. In one cation, the ring system is nearly planar, but in the other it is folded in the boat conformation as was also found for NiBr$_2$(acacH)$_2$.[219] Bond lengths and angles for the puckered cation are shown in Fig. 43.

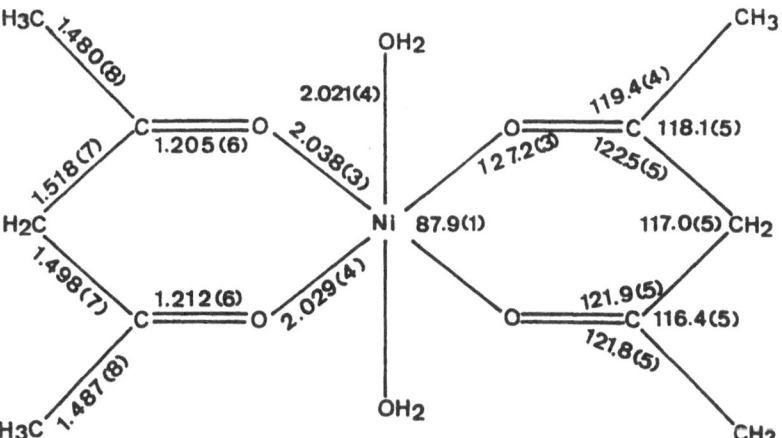

Fig. 43. Bond lengths (Å) and angles (°) for the puckered cation in *trans*-[Ni(acacH)$_2$(H$_2$O)$_2$](ClO$_4$)$_2$. Dihedral angles between the OCCO plane forming the bottom of the boat and the ONiO and CCC planes are 18.5° and 31.4°, respectively.[221]

5.1.4 O-Unidentate Coordination of Enol

Although MnBr$_2$(acacH)$_2$ has the same stoichiometry as NiBr$_2$(acacH)$_2$, their IR spectra are quite different (Table VIII). The bands observed at 1627 and 1564 cm^{-1} must be assigned to the C⋯O and C⋯C stretching vibrations of the enol molecule.[222]

X-Ray analysis[223] revealed that the manganese(II) complex is composed of infinite MnBr$_2$ chains with acacH enol molecules occupying the axial coordination sites to complete the octahedral geometry of *trans*-MnBr$_4$O$_2$. The acacH molecule is planar and coordinates to the metal atom through the carbonyl oxygen, holding the intramolecular hydrogen bond with the Mn—O and O⋯O′ distances being 2.20(2) and 2.56(3) Å, respectively.

Dimeric complexes [ReCl(CO)$_3$(β-dikH)]$_2$ were obtained in the reaction of chloropentacarbonylrhenium(I) with acacH or bzacH in refluxing benzene.[224] The molecular structure of the bzacH complex determined by X-ray diffraction[225] is shown in Fig. 44. The two planar enol molecules are approximately at 90° to each of the Re-Cl bridges and are on the same side of the ReCl$_2$Re plane.

Asymmetric internal hydrogen bonding in the coordinated bzacH is clearly indicated. The C=O, C—O, and O ⋯ O′ distances, 1.29(2), 1.37(2), and 2.52(2) Å, respectively, correspond well to those values (1.288(5), 1.310(4), and 2.489(5) Å, respectively) for free bzacH determined by neutron diffraction.[214]

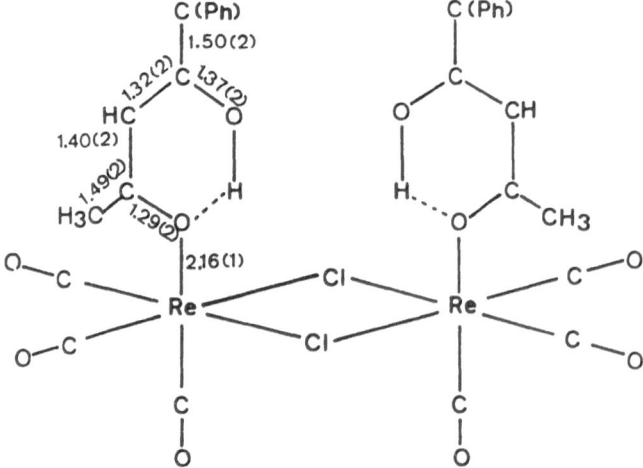

Fig. 44. Molecular structure of [ReCl(CO)$_3$(bzacH)]$_2$. The intramolecular O ⋯ O′ contact is 2.52(2) Å.[225]

5.1.5 $\eta^2(C, C')$ Coordination of Enol

A Pt(II) complex, K[PtCl(acac)(acac-C^3)] (28), which contains a central-carbon-bonded acac ligand,[226] reacts with a strong acid to give PtCl(acac)(acacH) (29).[227, 228] Lewis et al. investigated the structure of 29 by IR and NMR spectroscopy, and first proposed structure A in which an enol molecule of acacH coordinates to the metal in the $\eta^2(C, C')$ fashion,[227] but later favored structure B in which the enol molecule is bonded as a delocalized π-system:[228]

On the other hand, Behnke and Nakamoto supported structure A based on detailed infrared spectroscopic studies.[229] Preference for B by Gibson et al.[228] was based on the NMR equivalence of the methyl protons of the η-bonded acacH. However, Tsutsui et al.[230] favored structure A by showing that deprotonation of 29 to reproduce the carbon-bonded complex 28 is reversible and rapid on the NMR time scale in polar solvents, averaging the environments of methyl protons of the η^2-coordinated acacH.

5.2 Coordination Modes for Monoanions

5.2.1 O,O′-Chelation and Bridging

O,O′-Chelation is the most popular mode of coordination for β-dik anions,[231] and has been found for almost all of metallic[232] and metalloidal[233] elements. The oligomeric structures of some acac chelates of the first-row transition metals such as [Mn(acac)$_2$]$_3$,[234] [Fe(acac)$_2$]$_2$,[235] [Fe(acac)$_2$]$_4$,[236] [Co(acac)$_2$]$_4$,[237] [Ni(acac)$_2$]$_3$,[238, 239] and [Zn(acac)$_2$]$_3$[240] as well as [Cd(acac)$_2$]$_\infty$[241] are well documented,[242] in which one or both of the oxygen atoms of a chelating acac ligand are further linked to other metal atoms.

Figure 45 shows the molecular structure of [Mn(acac)$_2$]$_3$ as a very recent example.[234] Each Mn atom in the trimer is surrounded by six oxygen atoms, and one of the oxygen atoms of each acac ligand, O(j2), bridges the central and terminal Mn atoms. The average Mn—O distance in the central MnO$_6$ unit is 2.170(4) Å, while the six Mn—O distances in the terminal MnO$_6$ unit are grouped into two classes with averages Mn(1)—O(j1) = 2.099(2) Å and Mn(1)—O(j2) = 2.244(9) Å: thus, the Mn—O(bridging) distances are 0.145 Å longer than the Mn-O(nonbridging) distance. This large difference indicates that each acac ligand in this compound coordinates to the Mn atoms unsymmetrically.[234]

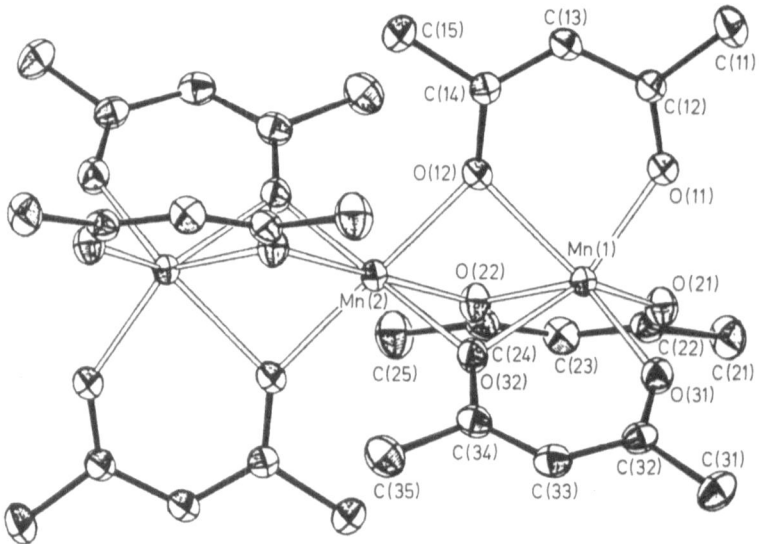

Fig. 45. Molecular structure of [Mn(acac)$_2$]$_3$.[234]

The coordination polyhedron of the terminal MnO_6 unit can be described as a near trigonal prism, while that of the central one is a near octahedron (trigonal antiprism), and they are fused to constitute a trimer by sharing triangular faces of oxygen atoms. This trimeric structure is remarkably different from that in $[Ni(acac)_2]_3$, where all the coordination polyhedra were described as octahedra,[238, 239] shown in Fig. 46.

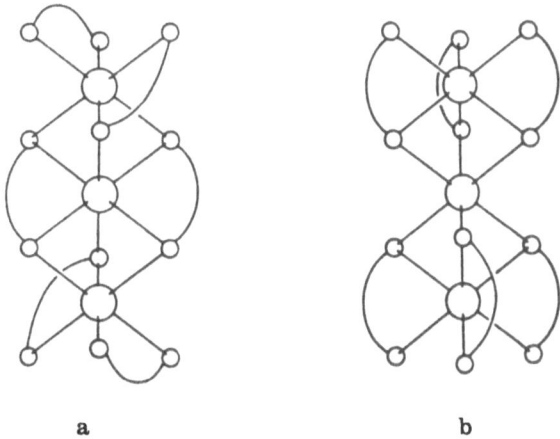

a   b

Fig. 46 a, b. Schematic drawing of the coordination in $[Ni(acac)_2]_3$ (**a**) and $[Mn(acac)_2]_3$ (**b**).[234]

Besides these self-associated binary compounds, there have been reported many mixed-ligand dinuclear complexes in which one of the oxygen atoms of the chelating β-dik ligand serves as the bridging donor atom. An example is the dimeric (allylamine)bis(acetylacetonato)-manganese(II), which was prepared by the equimolar reaction between allylamine and $Mn(acac)_2(H_2O)_2$ in ether.[243] As is seen in Fig. 47, the molecule is a centrosymmetric dimer, the Mn atom being surrounded octahedrally by five oxygen and one nitrogen atoms. Of the four oxygen atoms of two acac ligands chelating a Mn atom, one is further linked to the other Mn atom to complete its coordination octahedron.[243b]

The purely bridging mode of coordination has sometimes been inferred for β-dik anions and was confirmed by X-ray analysis for two dipivaloylmethanate (dpm) ions in $Er_8O(OH)_{12}(dpm)_{10}$.[244] The compound was obtained as a minor product during the recrystallization of $Er(dpm)_3$ from n-hexane, and sublimes with $Er(dpm)_3$ apparently without decomposition.

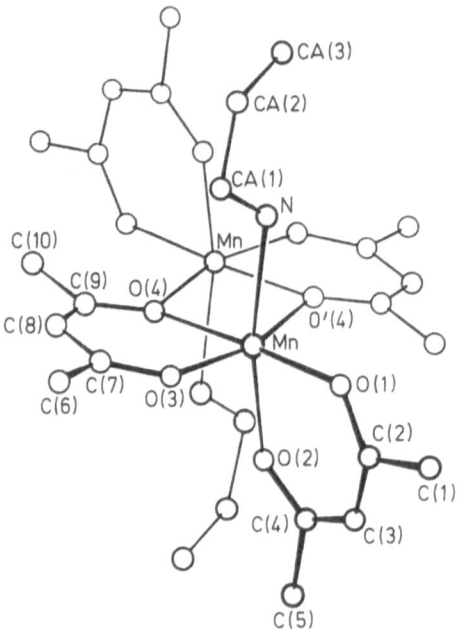

Fig. 47. Perspective drawing of the $[Mn(acac)_2(NH_2CH_2CH=CH_2)]_2$ molecule.[243b]

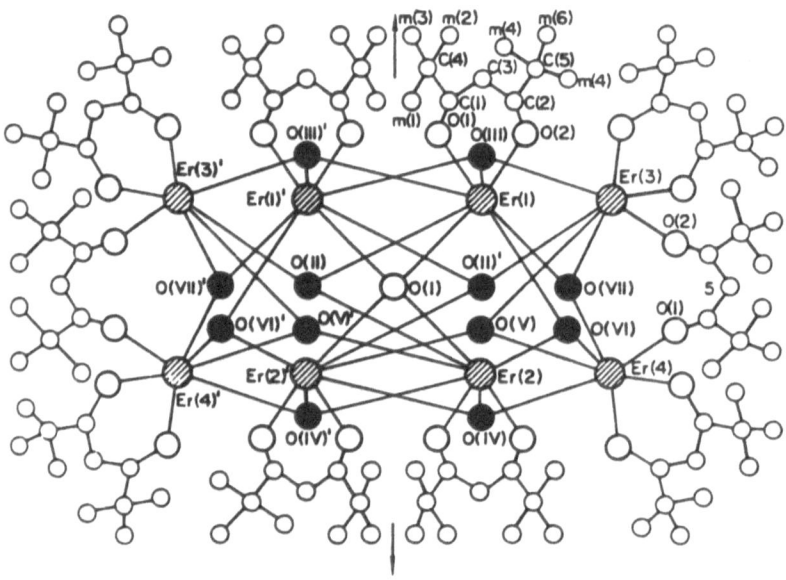

Fig. 48. Schematic drawing of the $Er_8O(dpm)_{10}(OH)_{12}$ molecule.[244]

As is shown by Fig. 48, the Er atoms are chelated by one dpm group each and cluster around a central oxygen atom. Each of the four Er atoms in the inner sphere constituting a distorted tetrahedron around the oxygen atom is bonded to five hydroxyl groups and is eight-coordinate. Four Er atoms in the outer sphere which formally complete dodecahedron around the central oxygen are bonded to four hydroxyl groups each and are seven-coordinate. The seventh ligand atom belongs to a non-chelating dpm group which bridges two different Er atoms. Each hydroxyl group is bonded to three Er atoms.[244]

5.2.2 Central Carbon Bonding and C,O,O'-Bridging

The central carbon bonding of β-dik was first found by Swallow and Truter in $Me_3Pt(acac-C^3)(bpy)$.[245] $K[Pt(acac)_2Cl]$ was obtained by Werner in 1901 and supposed to involve an O-unidentate acac ligand,[246] but X-ray analysis[247] as well as IR and NMR studies[248] revealed that the unidentate acac is bound to Pt through the central carbon. Since then, various metal complexes containing the central-carbon-bonded β-dik ligands have been reported:[249] $K[PtX(\beta\text{-dik})(\beta\text{-dik-}C)]$ (β-dik = tfac and bzac; X = Cl and Br),[250] $fac\text{-}Mn(CO)_3(LL)(\beta\text{-dik-}C)$ (β-dik = tfac and hfac; LL = bpy, phen, and 2py),[251] $AuL(\beta\text{-dik-}C)(\beta$-dik = acac and bzac; L = PPh_3, $PEtPh_2$, PEt_3, $P(p\text{-tolyl})_3$, and $AsPh_3$),[252] $Hg(\beta\text{-dik-}C)_2$ (β-dik = acac and dpm),[253] $Ir(acac)_2(acac-C^3)L$ (L = NH_3, py, n-$PrNH_2$, and $p\text{-}MeC_6H_4NH_2$),[254] and $(\eta\text{-}C_5Me_5)Ir(acac)(acac-C^3)$.[255]

Although tetraacetylethane,[209] 3,3'-dithiobis(2,4-pentanedione),[210] and 3,3'-trithiobis(2,4-pentanedione)[211] exist in the dienolic form, and the carbon-bonded hfac in $Mn(CO)_3(py)_2(hfac-C^3)$ was suspected to be enol,[251] the other compounds cited above contain β-dik of the keto form. Then, the two oxygen atoms of the carbon-bonded β-dik ligand should be able to coordinate with another metal atom. In fact $[Me_3Pt(\beta\text{-dik})]_2$ (β-dik = $C_3H_7COCHCOC_3H_7$[256] and etac[257]) were the first examples of dinuclear complexes in which the β-dik anion serves as a C,O,O'-bridging ligand. Trinuclear complexes $M[PtCl(acac)(acac-C^3)]_2$ (M(II) = Mn, Fe, Co, Ni, Cu, Zn, Cd, Pd, VO, and UO_2) and the like were prepared[258] and characterized by IR[259] and NMR[258] spectroscopy. The C,O,O'-bridging of β-dik in $Ir_2(acac)_6$[254] and $[(\eta\text{-}C_5Me_5)Rh(\beta\text{-dik})]_2[PF_6]_2$ (β-dik = acac, etac, and $CH_3COCHCOCH_2CH_2CH_3$)[255] was deduced from IR and NMR data and confirmed by X-ray analysis for $[(\eta\text{-}C_5Me_5)Rh(acac)]_2[BF_4]_2$.[255]

Bis(acetylacetonato)palladium(II) was found by Baba et al.[260] to react easily with Lewis bases (L) such as PPh_3, py, and Et_2NH to afford

Pd(acac)(acac-C^3)L in about 90% yields:

$$\tag{9}$$

These were the first examples of carbon-bonded complexes of acac with Pd(II); their molecular structures were confirmed by X-ray analysis for L = PPh$_3$[261] and Et$_2$NH.[262]

This kind of reaction was successfully applied to Pt(acac)$_2$, Me$_2$Au(acac), and Pd(hfac)$_2$ to yield Pt(acac)(acac-C^3)L (L = PPh$_3$,[263] PCy$_3$,[263] and py[264]), Pt(acac-C^3)$_2$(py)$_2$,[264] Me$_2$Au(acac-C^3)-(PMe$_2$Ph),[265] and Pd(hfac)(hfac-C^3)L (L = Me$_2$NH and other bases),[266] respectively.

The reaction of etacH with Na$_2$[PdCl$_4$] in aqueous alkaline solution followed by treatment of the product with an excess amount of nitrogen bases (L) in appropriate solvents gave the central-carbon-bonded complexes Pd(etac-C^2)$_2$L$_2$ (L = py, 2-Me-py, PhCH$_2$NH$_2$, and n-BuNH$_2$).[267] The molecular structure of trans-Pd(etac-C^2)$_2$(2-Me-py)$_2$ was determined by X-ray analysis.[268]

The reactions of binary and mixed-ligand bis(β-diketonato)-palladium(II) and -platinum(II) chelates with various nitrogen bases[269 – 271] and tertiary phosphines[272] have been studied extensively. Many types of complexes have been obtained depending on the nature of the central metals, β-dik ligands, and the attacking Lewis bases; the possible reaction routes are shown in Scheme 5.

For example, when a mixture of Pd(acac)$_2$ (30a) and Et$_2$NH was warmed to result in a clear solution and then kept in a refrigerator overnight, [Pd(acac)(Et$_2$NH)$_2$](acac) (33a) was obtained in a 66% yield.[269] On the contrary, when the solution was kept at room temperature to let the excess amine evaporate spontaneously, Pd(acac)(acac-C^3)(Et$_2$NH) (36a) was left in an 81% yield.[260]

When complex 30a was allowed to react with excess Et$_2$NH in solution, the absorption spectrum changed with time, exhibiting distinct isosbestic points to reach the spectrum of authentic 33a:[273]

$$Pd(acac)_2 + 2\,Et_2NH \rightarrow [Pd(acac)(Et_2NH)_2](acac) \tag{10}$$
$$\quad 30a \qquad\qquad\qquad\qquad\qquad 33a$$

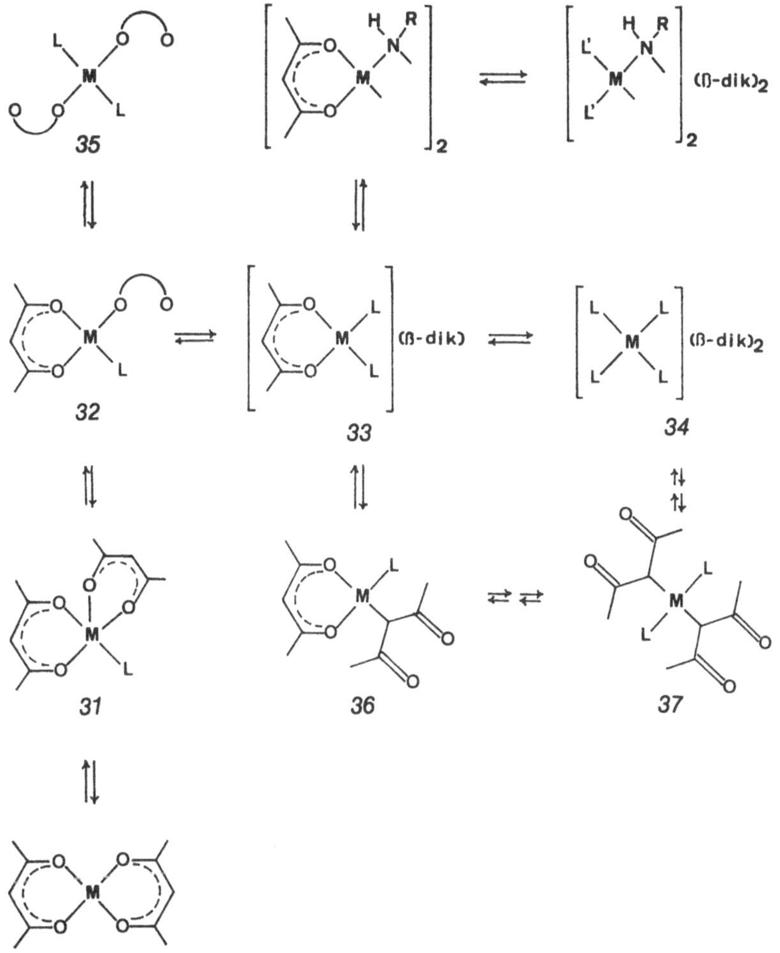

Scheme 5: Possible pathways for the reactions of bis(β-diketonato)-palladium(II) and -platinum(II) chelates with various Lewis bases.[194]

The second order rate constant of this reaction at 25 °C was 8.27×10^{-2}, 1.14×10^{-3}, and 9.4×10^{-4} dm^3 mol^{-1} s^{-1} in methanol, tetrahydrofuran, and benzene, respectively. Complex *36 a* also reacted with Et$_2$NH to afford *33 a* [Eq. (11)], but the rate (9.70×10^{-4} dm^3 mol^{-1} s^{-1} in methanol at 25 °C) was two orders of magnitude smaller than that for the overall reaction:[273]

$$\text{Pd(acac)(acac-}C^3)(\text{Et}_2\text{NH}) + \text{Et}_2\text{NH} \rightleftarrows [\text{Pd(acac)(Et}_2\text{NH})_2](\text{acac})$$

$$\text{36 a} \qquad\qquad\qquad\qquad\qquad\qquad \text{33 a} \qquad\qquad (11)$$

These preparative and kinetic results indicate that complex *36 a* is not produced directly from *30a* as an intermediate for *33a*, but is derived conversely via *33 a*. The equilibrium constant of Eq. (10) was determined to be 3.93×10^6 and $32.9 \text{ dm}^6 \text{ mol}^{-2}$ in methanol and dicloromethane, respectively, at 25°C. In methanol, the salt-like complex *33 a* is quite stable, and complex *36 a* is not involved at all in the equilibrium mixture. In dichloromethane, on the other hand, the equilibrium constant of Eq. (11) is $0.224 \text{ dm}^3 \text{ mol}^{-1}$ at 25°C and *33 a* is readily converted into *36 a*.[273]

In the case of the two-coordinate Hg(II) complex of 1,1,1,2,2,3,3-heptafluoro-7,7-dimethyl-4,6-octanedione, an intramolecular mechanism was proposed by Fish[274] for the linkage isomerization between the O-unidentate and C-unidentate states. In the present case, the rearrangement of the chelating acac to the central-carbon-bonded state induced by the reaction of Et_2NH with *30a* occurs by the following consecutive replacement mechanism:

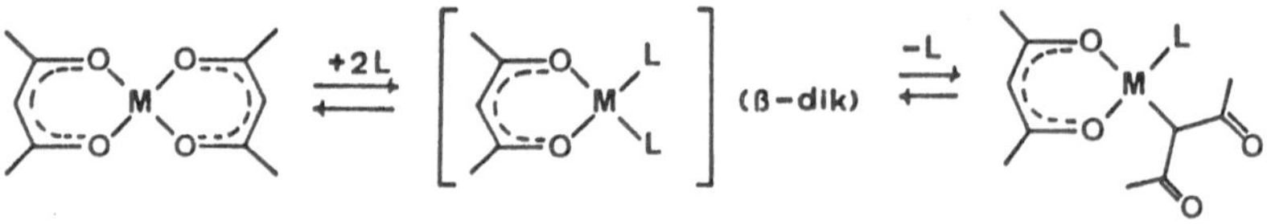

The nucleophilic attack of β-dik as a carbanion has been demonstrated not only on the unsaturared ligands such as olefin,[275–280] imine,[281] and cyanide,[282, 283] but also on the metal atoms, for example, in $Me_3PtI(bpy)$,[284] $PtBr_2(1,5\text{-hexadiene})$,[285] $Mn(CO)_4(LL)$ (LL = bpy, phen, and 2py),[251] and $(\eta\text{-}C_5Me_5)IrCl(acac)$.[255] In the presence of appropriate ancillary ligands, metal atoms prefer the carbon bonding to the oxygen bonding when they accept a β-dik anion as a unidentate ligand.

The tendency of a β-dik anion to form the central carbon bond is parallel with the preference of the keto tautomer over enol.[200–202] In the reaction with excess pyridine, both the chelating anions of ethyl acetonacetate in $Pd(etac)_2$ are transformed into the central-carbon-bonded state to result in $Pd(etac\text{-}C^2)_2(py)_2$, whereas $Pd(acac)_2$ yields only $Pd(acac)(acac\text{-}C^3)(py)$ with one acac chelate remaining intact.[269] Ito and Yamamoto[286] examined the ligand substitution reactions of $Pt(acac)(acac\text{-}C^3)(PPh_3)$ with several β-dikH in refluxing toluene and found that the keto-favoring β-dikH selectively replaces the carbon-bonded acac, while the enol-favoring β-dikH substitutes the chelating acac.

However, even the hfac anion can realize the central carbon bonding in some cases. Thus, the $Pd(hfac)(hfac\text{-}C^3)L$ complexes with 2,6-

dimethyl-, 3,5-dimethyl-, and 2,4,6-trimethylanilines[270] as well as di-methylamine, 2,6-dimethylpyridine, phenoxathiin, and phenazine[266] as L have been prepared. Recently, *cis*- and *trans*-$PtCl_2(Et_2S)_2$ were reported to react with $Tl(\beta\text{-dik})$ in methanol at room temperature to afford $Pt(\beta\text{-dik})(\beta\text{-dik-}C^3)(Et_2S)$ (β-dik = acac, tfac, and hfac).[287]

5.2.3 Outer-Sphere Coordination

Readiness of the reaction between $M(\beta\text{-dik})_2$ and Lewis bases (L) depends upon the basicities of both the β-dik anion and L, being in the sequence hfac > tfac > bzac $\simeq$ acac and alkylamines > benzyl-amine > pyridine. Reactions of $M(acac)(\beta\text{-dik})$ (β-dik = tfac and hfac) with L to afford complexes *33* (Scheme 5) demonstrate this trend, displacing tfac and hfac preferentially into the outer sphere. Although pyridine reacts with $Pt(tfac)_2$ to give $[Pt(py)_4](tfac)_2$, the less basic 4-cyanopyridine cannot displace tfac, but removes hfac in $Pt(hfac)_2$.[271]

Similarly, aniline and its derivatives, which are less basic than alkylamines and pyridine, cannot substitute β-dik such as acac, bzac, and tfac in $Pd(\beta\text{-dik})_2$ at room temperature, but reacts with $Pd(hfac)_2$ to give complexes *36*, *33*, and *34* depending on the reactants' mole ratio.[270] In refluxing benzene, however, anilines react with various $Pd(\beta\text{-dik})_2$ to give rise to the anilide-bridged dinuclear complexes $[Pd(\beta\text{-dik})\text{-}(\text{anilido})]_2$.[270]

The steric demand of Lewis base ligands L also exerts an important influence on the relative stabilities of the mixed ligand complexes. In general, primary amines and pyridine displace both of the β-dik ligands in $M(\beta\text{-dik})_2$ to afford complexes *34*, whereas secondary amines give only complexes *33*. Tertiary amines do not react with $M(\beta\text{-dik})_2$ to yield complexes *33*. Tribenzylamine and 2,6-diphenylpyridine did react with $Pd(hfac)_2$, but the products were orthometallated complexes.[269] Siedle[288] succeeded in characterizing *in situ* at lower temperatures the intermediates containing a unidentate hfac ligand in the orthometalla-tion reactions of $Pd(hfac)_2$ with N,N-dimethylbenzylamine and methyl benzyl sulfide.

A large number of the complexes *33* and *34* have been prepared.[269–272, 289, 290] In spite of their salt-like compositions, most of them are not soluble in water, but dissolve in aprotic solvents retaining the tight ion-pair structures as evidenced by molecular-weight determi-nations. The crystal structure of $[Pd(acac)(Et_2NH)_2](acac)$ *(33a)* at $-170\,^\circ$C determined by Kasai et al.[291] exhibits that the acac plane in the outer sphere stands aside of the coordination plane shearing a common C_2 axis with the dihedral angle of 80.8°. The cation and the anion are bound together by hydrogen bonds between the coordinated amine molecules and the carbonyl oxygens of the acac anion in the outer sphere, and the interaction persists in solution. The ^{1}H NMR data give the relative strength of the hydrogen bonds in the sequence bzac > acac > tfac > hfac[269] in accordance with the basicity sequence of β-dik anions.[292–294] The amine protons are deuterated by CDCl$_3$ and the rate is also parallel with the basicity of β-dik in the outer sphere. A mechanism assuming the alkylamido complex as an intermediate was proposed.[295]

5.2.4 O-Unidentate Linkage

Since the carbon-bonded complex 36 does not intervene between complexes 30 and 33, but is derived conversely via 33, a type-32 complex must be involved as an intermediate in the reaction pathway from 30 to 33. In fact, NMR monitoring disclosed that bis(trifluoroacetyl-acetonato)platinum(II), Pt(tfac)$_2$ reacts with an equimolar amount of PPh$_3$ to produce Pt(tfac)(tfac-O)(PPh$_3$) exclusively, which is converted into [Pt(tfac)(PPh$_3$)$_2$](tfac) by the reaction with another equivalent of PPh$_3$.[272]

Although the above O-bonded complex was not isolated and Pd(acac)(acac-O)(Et$_2$NH), the expected intermediate in Eq. (10), could not even be detected spectroscopically,[273] rather many compounds containing the β-dik anion as an O-unidentate ligand have been prepared. The first examples were Me$_2$Si(acac-O)$_2$ and Me$_3$Si(acac-O)[296] followed by Hg(β-dik-O)$_2$ (β-dik = acac, dpm, and 2,6-dimethyl-3,5-heptanedionate),[297] Pt(acac-O)$_2$L$_2$ (L = PEt$_3$[298] and piperidine[299]), Pt(acac)(acac-O)P(o-tolyl)$_3$,[272]   Cu(acac)(hfac-O)(phen),[300]   and M(CO)$_5${PPh(acac-O)$_2$} (M = Cr and W).[301]

The central carbon bonding of β-dik is more favorable for Pd(II) than Pt(II) and the reverse is true as preference for the O-unidentate linkage. Thus, Pd(acac)$_2$ reacts with PEt$_3$ to give Pd(acac)(acac-C^3)(PEt$_3$) exclusively,[272] while Pt(acac)$_2$ affords only Pt(acac-O)$_2$(PEt$_3$)$_2$.[298]

On the other hand, the reaction of piperidine with Pt(acac)$_2$ gives a pair of linkage isomers Pt(acac-O)$_2$(piperidine)$_2$ and [Pt(acac)(piperidine)$_2$](acac), whereas Pt(tfac)$_2$ and Pt(hfac)$_2$ produce only 35 and 33, respectively.[271] Preference for the O-unidentate bonding by β-dik is thus in the sequence: tfac > acac > hfac.

The O-unidentate acac anions involved in the trialkylsilyl compounds are a mixture of cis and trans isomers with respect to the C=C bond, and the cis isomer undergoes rapid intramolecular head-to-tail rearrangement, whereas the trans isomer is stereochemically rigid.[302] On the contrary, the acac anions involved in the above-mentioned Pt(II) complexes have the *cis* configuration exclusively, but show no fluxional motion at room temperature.[271, 298] The *cis* configuration of the unidentate β-dik ligands in Cu(acac)(hfac-*O*)(phen)[300] and Cr(CO)$_5${PPh(acac-*O*)$_2$}[301] was confirmed by X-ray analysis.

The O-unidentate bonding of tfac seems to be much more stable than that of symmetric β-ketoenolates and a number of complexes have been obtained:[271, 272] M(tfac)(tfac-*O*)L (M = Pd and Pt, L = P(*o*-tolyl)$_3$; M = Pd, L = PCy$_3$), M(acac)(tfac-*O*)L (M = Pd and Pt, L = P(*o*-tolyl)$_3$; M = Pt, L = PPh$_3$ and PEt$_3$), Pt(tfac-*O*)$_2$L$_2$ (L = PEt$_3$, PCy$_3$, Et$_2$NH, and piperidine), and Pt(acac-*O*)(tfac-*O*)(PEt$_3$)$_2$.

Kinetic and equilibrium studies[303] showed that the reaction:

$$\text{Pd(tfac)}_2 + \text{P(\textit{o}-tolyl)}_3 \rightarrow \text{Pd(tfac)(tfac-\textit{O})\{P(\textit{o}-tolyl)}_3\}$$

is reversible and the equilibrium constant was obtained as 1.38×10^3, 4.35×10^3, and larger than 10^9 dm^3 mol^{-1} at 25°C in benzene, dichloromethane, and methanol, respectively. The steric requirement of P(*o*-tolyl)$_3$ seems to prevent formation of [Pd(tfac){P(*o*-tolyl)$_3$}$_2$](tfac) and Pd(tfac)(tfac-*C*^{3}){P(*o*-tolyl)$_3$}.

The ^{1}H and ^{13}C NMR data indicate that the unidentate tfac anions in these complexes are bound to the metal atom through the acetyl

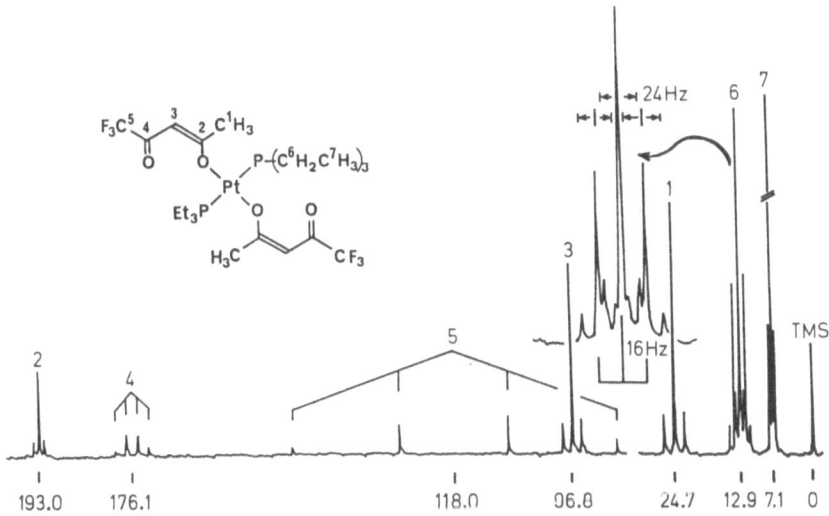

Fig. 49. ^{13}C{^{1}H} NMR spectrum at 15.04 MHz of Pt(tfac-*O*)$_2$(PEt$_3$)$_2$ in CD$_2$Cl$_2$. The J(Pt—C) values are 55, 28, and 55 Hz for CH$_3$, CO, and CH of tfac, and 24 and 15 Hz for CH$_2$ and CH$_3$ of PEt$_3$, respectively. ^{1}J(C—F) = 292 Hz and ^{2}J(C—F) = 31 Hz.[272]

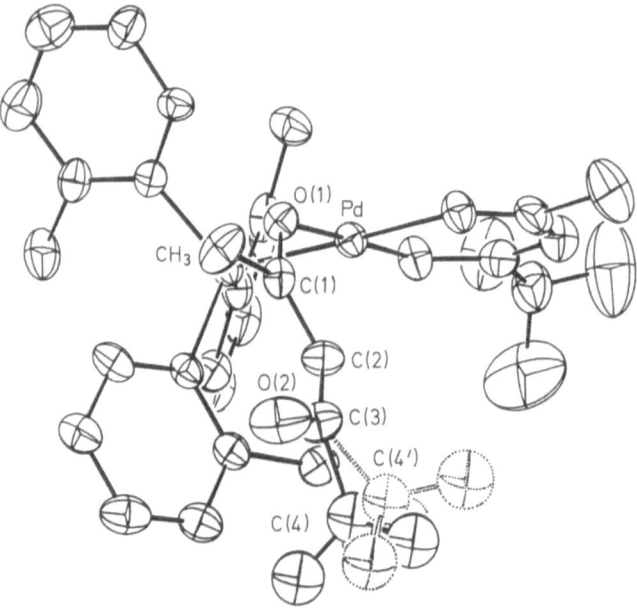

Fig. 50. ORTEP drawing of Pd(tfac)(tfac-O){P(o-tolyl)₃} involving the disordered CF₃ group.[304]

oxygen.[272] For example, the $^{13}C\{^1H\}$ spectrum of Pt(tfac-O)₂(PEt₃)₂ shown in Fig. 49 is compatible with the proposed structure.

X-Ray analysis of Pd(tfac)(tfac-O){P(o-tolyl)₃} confirmed the acetyl-oxygen bonding of the unidentate tfac to Pd.[304] As is seen in Fig. 50, the dangling CF₃CO group is positioned *trans* to the bonding oxygen atom around the C=C bond, the torsion angle between the C(1)-CH₃ and C(3)-O(2) bonds being 1.1°.[304] This geometrical structure is not consistent with the NMR data, indicating that the O-unidentate tfac ligand has primarily the *cis* configuration in CDCl₃.[272]

Recently, palladium(II) complexes of the Pd(β-dik-O)(pyridyl)(PEt₃)₂ type containing an O-unidentate acac, tfac, or hfac anion and a 2-, 3-, 4-pyridyl or 6-chloro-2-pyridyl group as ligands were prepared and characterized mainly by ^1H and ^{13}C NMR spectroscopy.[305] The acac and hfac ligands in the Pd(β-dik-O)(C₅H₃(6-Cl)N-C²)(PEt₃)₂ complexes undergo the head-to-tail intramolecular donor-atom exchange:

The NMR studies indicate that the dangling acetyl group of the acac ligand is positioned *cis* to the coordinating oxygen atom with respect to the C=C bond. As a mechanism for Eq. (12) a simple oscillatory motion of the β-dik ligand spanning the apical and basal coordination sites in the square-pyramidal intermediate is proposed:[305]

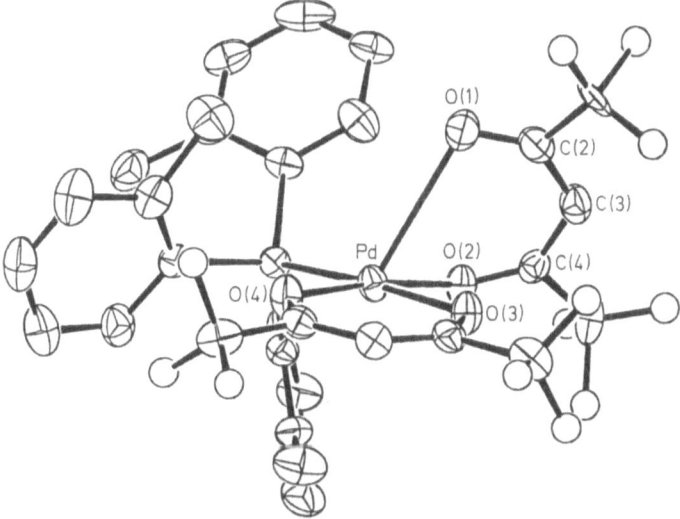

Five-coordinate complexes of d^8 metals are believed to be involved as intermediates in ligand substitution[306] and some *cis-trans* isomerization[307] reactions, but they have scarcely been isolated. The reactions of Pd(hfac)$_2$ with an equivalent amount of P(o-tolyl)$_3$ in *n*-hexane at room temperature gave Pd(hfac)$_2${P(o-tolyl)$_3$} *(31a)* in an 85% yield. Similarly, Pt(hfac)$_2${P(o-tolyl)$_3$} *(31b)* and Pt(hfac)$_2$(PCy$_3$) *(31c)* were also obtained in 89 and 93% yields, respectively.

The molecular structures of *31a* and *31c* were determined by X-ray analysis[308] and the former is shown in Fig. 51. The Pd atom is five-

Fig. 51. Molecular structure of Pd(hfac)$_2${P(o-tolyl)$_3$}.[308]

coordinate and has a very distorted square-pyramidal coordination, one of the two hfac anions spanning the apical and basal coordination sites. The apical atom [O(1)] largely deviates from the ideal site for tetragonal symmetry, the M—O(1) bond intersecting the basal plane defined by the O(2), O(3), O(4), and P atoms at 65.3(2)°. The apical M—O bond (2.796(6) Å) is much longer than the basal ones (average 2.034(6) Å), indicating the weaker interaction between M and the apical O.[308]

^{1}H, ^{13}C, and ^{19}F NMR studies indicate that these complexes *(31a–31c)* are stereochemically rigid in CD$_2$Cl$_2$ at -50°C but are fluxional at room temperature. Two kind of twist modes were proposed as the mechanism for the intramolecular coordination-site exchange.[308]

An analogous compound Pd(hfac)$_2$(PPh$_3$)[309] and the complex cations in

[Pd(hfac)(triphos)] [BPh$_4$]
(triphos = Ph$_2$PCH$_2$CH$_2$P(Ph)CH$_2$CH$_2$PPh$_2$)[310]

were also shown by X-ray analysis to have similar distorted square-pyramid structures. They are also fluxional in solution. The activation parameters for the coordination-site exchange reactions of Pd(hfac)$_2${P(*o*-tolyl)$_3$} and [Pd(hfac)(triphos)]$^+$ measured by ^{19}F NMR line shape analysis are $\Delta H^\ddagger = 7.7 \pm 1$ and 8.1 ± 1 kcal mol^{-1}, and $\Delta S^\ddagger = -20 \pm 5$ and -21 ± 3 cal deg^{-1} mol^{-1}, respectively.[309, 310]

5.2.5 η-Allylic Coordination

Ethanolysis of diketene to produce ethyl acetoacetate (etacH) is catalyzed by [PdCl$_4$]$^{2-}$, and yellow crystals separated at the end of the reaction were identified as [PdCl(η^3-etac)]$_2$ *(38)* in which the monoanion of ethyl acetoacetate coordinates to Pd(II) in an η-allylic fashion.[311] This complex is also prepared by the direct reaction between PdCl$_2$ and etacH in water at 70°C in an 85% yield,[312] and its structure was confirmed by X-ray crystallography.[313]

The molecular structure of *38* resembles that of [PdCl(η-allyl)]$_2$,[314] the dihedral angle between the Pd(μ-Cl)$_2$Pd plane and the η^3-etac plane being 108°.[313] The dimeric molecule in a crystal is connected with adjacent molecules by the intermolecular hydrogen bonds. In a CDCl$_3$ solution, the intermolecular hydrogen bonds are broken and instead the intramolecular hydrogen bond is maintained. The ^{1}H NMR spectrum shown in Fig. 52 is well interpreted based on the proposed structure.[312]

The reaction of PdCl$_2$(PhCN)$_2$ with acacH in acetone gave [PdCl(acac)]$_2$ at 0°C and [PdCl(η^3-acac)]$_2$ *(39)* at room temperature.

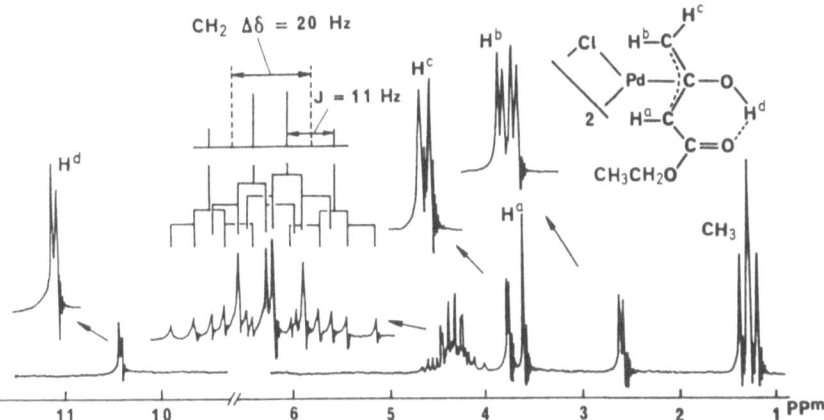

Fig. 52. ^{1}H NMR spectrum of [PdCl(η^3-etac)]$_2$ in CDCl$_3$. J(bc) = 3.4 Hz, J(bd) = 1.2 Hz, and J(CH$_3$—CH$_2$) = 9.2 Hz.[312]

By means of the bridge-splitting reactions, these insoluble compounds were converted into the soluble mononuclear ones, PdCl(acac)L and PdCl(η^3-acac)L (L = PPh$_3$ and AsPh$_3$), respectively.[315] They constitute two pairs of linkage insomers of a novel type.

5.2.6 Terminal Carbon Bonding

Complex *38* reacts with py, bpy, and their derivatives in benzene to afford mononuclear complexes PdCl(CH$_2$COCH$_2$COOEt)L$_2$ containing the terminal-carbon-bonded β-ketoester.[316] The pyridine and 4-methylpyridine complexes are a mixture of *cis* and *trans* isomers in a ratio of 1:5 each, while 2-methyl- and 2,6-dimethyl-pyridine complexes are *trans* exclusively. The molecular structure of *cis*-PdCl(CH$_2$COCH$_2$COOCH$_2$Ph)(py)$_2$ was determined by X-ray.[317]

In a similar way, complex *39* reacts with bpy and other bidentate ligands in chloroform to give PdCl(CH$_2$COCH$_2$COCH$_3$)(bpy) *(40)* etc. Although the terminal-carbon-bonded β-ketoester in the corresponding complex PdCl(CH$_2$COCH$_2$COOEt)(bpy) *(41)* is keto exclu-

40–keto   **40–enol**

sively, complex *40* is a mixture of keto and enol tautomers with the equilibrium quotient $K = [\text{enol}]/[\text{keto}] = 0.7$ in $CDCl_3$ at 25 °C.[315]

Unsubstituted acacH favors the enol form, K being 6.7 in chloroform at 33 °C.[318] The present result indicates that the {(bpy)ClPd} moiety in *40* serves as an electron-releasing substituent on acacH.

5.3 Coordination Modes for Dianions

There are seven coordination modes for dianions of β-dicarbonyl compounds, which are included in Fig. 41. Most of them were found rather recently.

5.3.1 Central Carbon Bonding

3,3-Di-μ-selenobis(2,4-pentanedione) was prepared by reaction of $SeCl_4$ with acacH, and its diselenacyclobutane structure with four acetyl groups was inferred from the IR and NMR spectra; the $ν(CO)$ band at 1706 cm^{-1} and a single proton NMR signal were observed.[319]

5.3.2 Chelation through Terminal Carbons

Dichloro(2,4-pentanedionato(2-)-C^1,C^5)tellurium(IV) was obtained from the reaction of $TeCl_4$ with acacH and reduced with aqueous $NaHSO_3$ to afford 2,4-pentanedionato(2-)-C^1,C^5-tellurium(II). The C,C-chelated structures of these complexes were deduced on the basis of spectroscopic data[319] and ascertained by X-ray analysis.[320, 321] The molecular structures of 1,3-dimethyl-, 1,5-dimethyl-, 3,3-dimethyl-, and 1,3,5-trimethyl-2,4-pentanedionato(2-)-C^1,C^5-tellurium(II) were also determined.[322 – 325]

5.3.3 Dienediolate Chelation

Dicarbonyl(η-cyclopentadienyl)(phenyldichlorophosphane)manganese reacted with neat acacH in the presence of Et_3N to give (η-C_5H_5)(CO)$_2$Mn{Ph$\overline{POC(=CH_2)CH=C(CH_3)O}$}, whose molecular structure was determined by X-ray to reveal that the acac(2-) ion forms the dienediolate chelate with the phosphorus atom.[326] Similarly, reactions of (CO)$_5$M(PPhCl$_2$) with acacH in the presence of Et_3N gave (CO)$_5$M{Ph$\overline{POC(=CH_2)CH=C(CH_3)O}$} (M = Cr and W), and the molecular structure of the Cr(0) complex was determined by X-ray analysis.[301]

Very recently, Kemmitt et al.[327] obtained a first example of a transition metal complex in which the β-dicarbonyl ligand is chelating as a dienediolate dianion. Thus, 1,5-diphenyl-1,3,5-pentanetrione reacted with $PtCO_3(PPh_3)_2$ in warm ethanol to afford $Pt\{OC(CHCOPh)CHC(Ph)O\}(PPh_3)_2$ which was characterized by X-ray crystallography (Fig. 53).

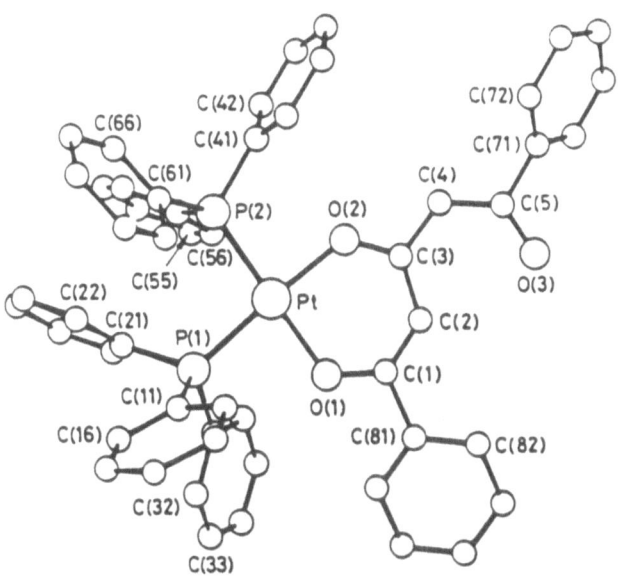

Fig. 53. Molecular structure of $Pt\{OC(CHCOPh)CHC(Ph)O\}(PPh_3)_2$. Selected bond lengths (Å): Pt—O(1) 2.033(10), Pt—O(2) 2.026(6), O(1)—C(1) 1.305(16), O(2)—C(3) 1.296(17), C(1)—C(2) 1.355(14), C(2)—C(3) 1.427(19), C(1)—C(81) 1.503(16), C(3)—C(4) 1.434(14), C(4)—C(5) 1.389(19), C(5)—O(3) 1.27(2), C(5)—(C71) 1.497(13).[327]

The six-membered chelate ring is essentially planar, and shows the C—C and C—O bond lengths intermediate between those of single and double bonds, indicating the usual electrom delocalization. The CHC(Ph)O substituent is also planar and the C—C and C—O bond distances are indicative of a highly conjugated system. The angle between the C(4)—C(5)—O(3)—C(71) and O(2)—C(3)—C(2)—C(1)—O(1) planes is 8.9°.[327]

5.3.4 C,O,O'-Bridging

As was noted previously (p. 97), $PdCl(CH_2COCH_2COCH_3)(bpy)$ (*40*, HY) still has an ionizable hydrogen and reacted with $Cu(acac)_2$,

VO(acac)$_2$, and Pd(acac)$_2$ to afford insoluble trinuclear complexes CuY$_2$, VOY$_2$, and PdY$_2$. Similar reactions of *40* with excess Be(acac)$_2$, Pd(acac)$_2$, and Pd(tfac)$_2$ to replace only one of the chelating ligands gave soluble dinuclear complexes M(β-dik)Y. The ligand substitution reactions of [Cu(acac)(bpy)]ClO$_4$ and [Pd(hfac)(bpy)](hfac) with *40* also produced [CuY(bpy)]ClO$_4$ and [PdY(bpy)](hfac), respectively. These soluble complexes were characterized by NMR as well as IR spectroscopy. The acac(2-) ligand is bound to a Pd atom through the terminal carbon and chelates another metal atom with two oxygen atoms.[328]

5.3.5 η-Allylic Coordination

Complexes *40* and *41* were treated with Tl(acac) in benzene and dichloromethane, respectively, at room temperature. Contrary to expectation, the acac ion was not coordinated with palladium in place of the chloride ligand, but acted as a base to accept a proton from the terminal-carbon-bonded β-dik ligand in *40* and *41*, resulting in novel trihapto complexes of β-dik(2-) ions, Pd(η^3-acac(2-))(bpy) *(42)* and Pd(η^3-etac(2-))(bpy) *(43)* in 93 and 97% yields, respectively.[312]

As is indicated in Scheme 6, complex *43* was also derived from *38* via another route. The bridging chloride in *38* was first displaced by the reaction with AgClO$_4$ in acetone. Ligand bpy was then added to give

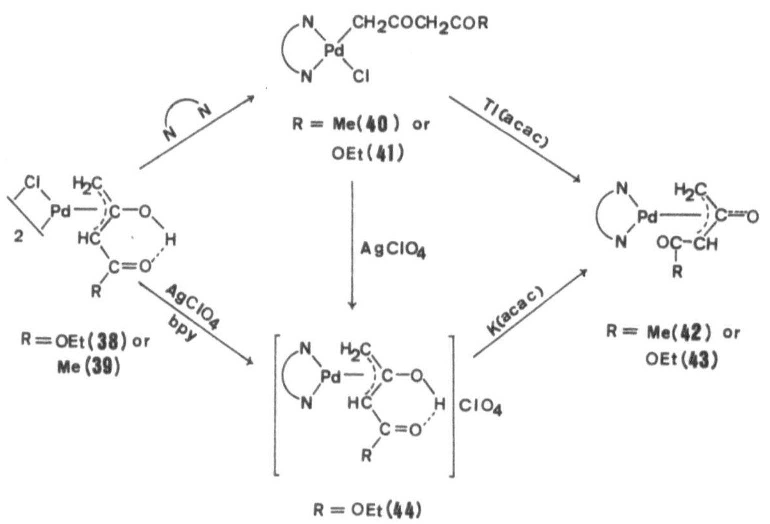

Scheme 6. Reaction steps for [PdCl(η^3-β-dik)]$_2$ to produce Pd(NN)(η^3-β-dik(2−)), where β-dik = acac or etac and NN = bpy, Me$_2$ bpy or phen.[312]

[Pd(η^3-etac)(bpy)]ClO$_4$ *(44)*, which retains the η^3 structure of the etac monoanion. Several bases were examined and K(acac) was the best in abstracting the enolic proton in *44* to produce complex *43*. The yield of *43* from *38* via this route was also excellent (95%).[312]

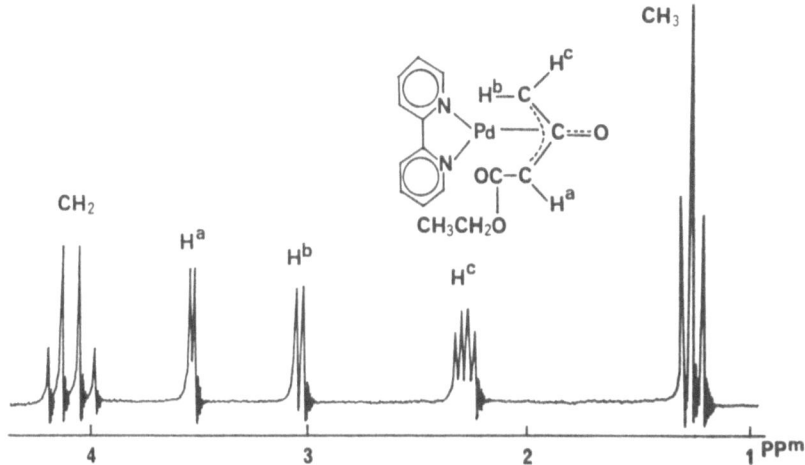

Fig. 54. ^{1}H NMR spectrum of Pd(η^3-etac(2−)) (bpy) in CD$_2$Cl$_2$. Signals from the bipyridine-ring protons are omitted. J(bc)=4.4 Hz and J(ac)=2.3 Hz.[312]

Figure 54 exemplifies the ^{1}H NMR spectrum of *43*, which is much simpler than that of *38* shown in Fig. 52. In the case of *38*, the ethoxycarbonyl group is held at the *syn* position by virtue of intra-molecular hydrogen bonding. The diastereotopic methylene protons exhibit the 16-line multiplet, while the enolic proton resonates at δ10.46 as a 1.2-Hz doublet owing to coupling to H^b in accordance with the so-called W-rule.[329] In complex *43*, the enolic proton is lost and the ethoxycarbonyl group now occupies the anti position, the ethyl CH$_2$ protons showing a simple quartet. Free rotation around the C—C bond may equalize the time-average environments of the methylene protons. On the other hand, the methine proton H^a appears as a doublet coupling to the syn H^c proton.[312]

A Pt(II) complex containing an acac(2-) ion as a trihapto ligand, Pt(η^3-acac(2-)){P(p-ClC$_6$H$_4$)$_3$}$_2$, was prepared in a 49% yield by the reaction of Pt(acac)$_2$ with tris(p-chlorophenyl)phosphine in chloroform under nitrogen at room temperature, and was characterized by ^{1}H, ^{13}C, and ^{31}P NMR spectroscopy.[330] It is surprising that this reaction proceeds under such very mild conditions. As is discussed shortly, a type-33 complex (Scheme 5) [Pt(PR$_3$)$_2$(acac)](acac) seems to be involved as an intermediate, and deprotonation of the chelating acac by the outer-sphere acac anion may result in the acac(2-) complex.

The η^3-acac(2-) complex is protonated conversely in the reaction with pyridinium perchlorate in a mixture of acetone and dichloromethane at 40°C to afford $[Pt(CH_2COCH_2COCH_3)(py)\{P(p\text{-}ClC_6H_4)_3\}_2]ClO_4$ in 64% yield.[330] It is noteworthy that the terminal-carbon-bonded acac in this complex is exclusively enol, indicating that the $\{(py)(PR_3)_2PtCH_2\}$ moiety serves as a strong electron-withdrawing group comparable to CF_3, although the $\{(bpy)ClPdCH_2\}$ group is electron releasing (p. 98).

Recently, a series of Pt(II) complexes Pt(CHRCOCHR)L$_2$ was prepared by the reactions of Pt(CO$_3$)L$_2$ with esters of 3-oxo-pentanedioic acid[331] or 2,4,6-heptanetrione,[327] where R = COOMe, L = PPh$_3$, AsPh$_3$, PMePh$_2$, PMe$_2$Ph, $\frac{1}{2}$(Ph$_2$PCH$_2$CH$_2$PPh$_2$); R = COOEt, L = PPh$_3$, AsPh$_3$; R = COO(n-Pr), L = PPh$_3$, AsPh$_3$. Similar Pd(II) complexes Pd(CHRCOCHR)L$_2$ (R = COOMe, L = PPh$_3$, PMePh$_2$, PMe$_2$Ph, PEt$_3$, AsPh$_3$, $\frac{1}{2}$(bpy); R = COOEt, L = PPh$_3$, AsPh$_3$, $\frac{1}{2}$(bpy); R = COO(n-Pr), L$_2$ = bpy) were also ob-tained from the reactions of [Pd$_2$(dba)$_3$] · CHCl$_3$ (dba = dibenzy-lideneacetone) with esters of 3-oxopentanedioic acid in diethyl ether in the presence of donor ligand L and dioxygen.[332]

X-Ray diffraction studies on

[Pt{CH(COOMe)COCH(COOMe)}(PPh$_3$)$_2$] · H$_2$O,[331]
Pt{CH(COMe)COCH(COMe)}(PPh$_3$)$_2$,[327]

and

[Pd{CH(COOMe)COCH(COOMe)}L$_2$] · H$_2$O
(L = PPh$_3$, AsPh$_3$, $\frac{1}{2}$(bpy)[332]

showed the presence of highly non-planar metallacylic rings and indicated that a bonding description of these complexes should include a contribution from the η-allylic structure M(η^3-CHRCOCHR)L$_2$.

5.3.6 C,O-Chelation

Bis(trifluoroacetylacetonato)platinum(II) reacted with PPh$_3$, P(p-ClC$_6$H$_4$)$_3$, and AsPh$_3$ in diethyl ether or chloroform at room tempera-ture to give the Pt(tfac(2-)-C,O)L$_2$ complexes in 55, 39, and 87% yields, respectively.[333] ^{1}H NMR monitoring revealed that when Pt(tfac)$_2$ was mixed with twice molar PPh$_3$ in CDCl$_3$, [Pt(tfac)(PPh$_3$)$_2$](tfac) was first produced and then converted into Pt(tfac(2-)-*C,O*)(PPh₃)₂.[272] The chelating tfac ligand may have been deprotonated by the tfac anion in the outer sphere.

Pd(tfac)₂ reacted readily with an equimolar amount of bpy in benzene at room temperature to yield [Pd(tfac)(bpy)](tfac).[269] When it was isolated and dissolved in hot dichloromethane, the type-*4* compound changed spontaneously to Pd(tfac(2-)-*C,O*)(bpy).[334] The reaction of Pd(tfac)₂ with twice molar PPh₃ in CDCl₃ was also confirmed by ¹H NMR spectroscopy to form [Pd(tfac)(PPh₃)₂](tfac).[272] Contrary to the Pt(II) case, however, the Pd(II) compound is stable and the proton transfer reaction does not occur spontaneously. The powerful and noncoordinating base 1,8-bis(dimethylamino)naphthalene (proton sponge) is effective for deprotonation, giving rise to Pd(tfac(2-)-*C,O*)(PPh₃)₂. Employment of pyridine and its derivatives as a base effected substitution of one PPh₃ molecule besides deprotonation of the chelating tfac to result in Pd(tfac(2-)-*C,O*)(PPh₃)L.

The molecular structure of Pd(tfac(2-)-*C,O*)(PPh₃)(2,6-Me₂py) was determined by X-ray diffraction and is shown in Fig. 55. The Pd(II) atom

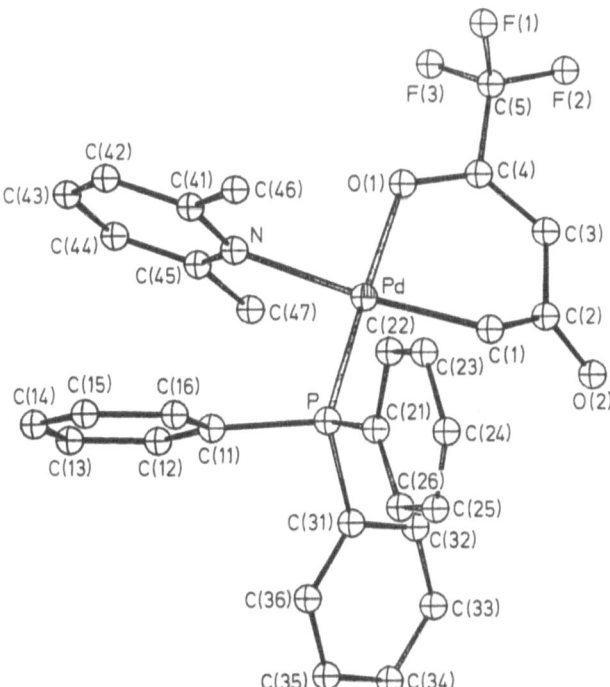

Fig. 55. Molecular structure of [Pd(tfac(2−)-*C,O*)(PPh₃)(2,6-Me₂py)] · C₆H₆.[334]

is C,O-chelated by the tfac dianion as was expected from the ^{1}H, ^{13}C, ^{19}F, and ^{31}P NMR spectroscopy. The tfac(2-) anion is not planar and makes a distorted six-membered ring with the central Pd atom. The locally planar parts are formed by the C(1), C(2), C(3), and O(2) atoms and by the C(3), C(4), C(5), and O(1) atoms. The dihedral angle between these planes is 29.3(2)°.[334]

Deprotonation of the chelating β-dik ligand in [M(β-dik)L$_2$]$^+$ may produce a dienediolate chelate, which is not stable in these Pt(II) and Pd(II) cases and transformed into more suitable coordination modes. The tfac(2-) ion prefers the C,O-chelation both for Pt(II) and Pd(II), whereas the acac(2-) ion favors the trihapto coordination over the C,O-chelation in either case:

R = CH$_3$ or CF$_3$

In the case of Pt{OC(CHCOPh)CHC(Ph)O}(PPh$_3$)$_2$ (p. 99), the presence of phenyl groups in the dianion of triketone may prevent transformations of the above type, thus stabilizing the dienediolate structure.[327]

5.3.7 η^3,O,O′-Bridging

The trihapto dianionic ligands in *42* and *43* still have uncoordinated oxygen atoms, suggesting the possibility of interaction with another metal atom. In order to improve solubilities, the bipyridine ligand in *42* and *43* was replaced by two molecules of PPh$_3$ or diphosphine ligands (PP) such as dppe and *cis*-1,2-bis(diphenylphosphino)ethylene (dpe) to obtain Pd(η^3-acac(2-))(PP) and Pd(η^3-etac(2-))(PP), which were characterized mainly by ^{1}H, ^{13}C, and ^{31}P NMR spectroscopy. These complexes reacted with [Pd(PP)(H$_2$O)$_2$](ClO$_4$)$_2$ in methanol at 0°C to give [(PP)Pd(η^3-β-dik(2-)-O,O′)Pd(PP)](ClO$_4$)$_2$ complexes in good yields.[335]

Single crystals of these compounds suitable for X-ray studies have not been obtained, but their structures were inferred from the ^{1}H, ^{13}C, and ^{31}P NMR spectra. Figure 56 exemplifies the ^{13}C NMR signals from the

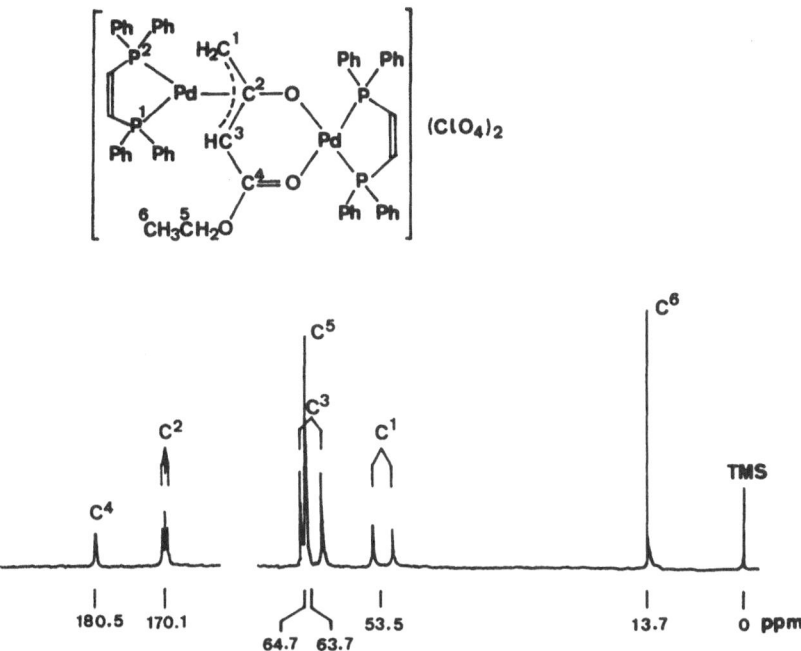

Fig. 56. $^{13}\text{C}\{^1\text{H}\}$ NMR signals at 15.04 MHz from the etac(2-) ligand in [(dpe)Pd(η^3-etac(2-)-O,O')Pd(dpe)](ClO$_4$)$_2$ in acetone-d_6. $J(\text{P}^1\text{—C}^1) = 43$ Hz, $J(\text{P}^1\text{—C}^2) = J(\text{P}^2\text{—C}^2) = 6$ Hz, and $J(\text{P}^2\text{—C}^3) = 33$ Hz.[335]

etac(2-) ligand in [(dpe)Pd(η^3-etac(2-)-O,O')Pd(dpe)](ClO$_4$)$_2$, which accord with the proposed dinuclear structure.[335]

5.4 η^3,C,O-Bridging of the Acetylacetonate Trianion

Very recently, Okeya et al. obtained a dinuclear Pt(II) complex containing an acac(3-) ion as the η^3,C,O-bridging ligand.[336] In hot methanol, Pt(acac)$_2$ reacted readily with a twice molar amount of PPh$_3$ to afford [Pt(acac)(PPh$_3$)$_2$](acac). When the reaction mixture was kept at 60°C for more than four hours, a novel dinuclear complex [Pt$_2$(C$_5$H$_5$O$_2$)(PPh$_3$)$_4$](acac) was produced together with a small amount of Pt(η^3-acac(2-))(PPh$_3$)$_2$. Addition of KPF$_6$ resulted in white crystals of [Pt$_2$(C$_5$H$_5$O$_2$)(PPh$_3$)$_4$]PF$_6$.

The ^{1}H, ^{13}C, ^{31}P, and ^{195}Pt NMR spectra for these compounds are consistent with the following structure proposed for the complex cation in which the acac(3-) anion coordinates with a Pt(II) atom in a trihapto fashion and C,O-chelates another Pt(II) atom:[336]

5.5 Concluding Remarks

Although β-dicarbonyl compounds are very classic as ligands, a number of new coordination modes of their mono- and di-anions have been explored only in recent years. Their preference for η^1 and η^3 carbon bonding to the usual O,O'-chelation has been demonstrated mostly in the Pd(II) and Pt(II) complexes. It is hoped, however, that many complexes of other metals containing β-dik ligands in nonclassical coordination states will be prepared in the future, and that the role of β-dik ligands in various metal complex-reactions will be elucidated. The acac(2-) and etac(2-) ions stabilized in the coordination sphere of Pd(II) may be useful for organic syntheses, since the acetoacetate dianion is one of the fundamental building blocks in the biogenesis of natural products.[337]

6 Abbreviations

acacH	2,4-pentanedione (acetylacetone)
acac, acac(2-), acac(3-)	mono-, di-, tri-anion of acetylacetone
bpy	2,2'-bipyridine
bzacH	1-phenyl-1,3-butanedione (benzoylacetone)
dbmH	1,3-diphenyl-1,3-propanedione (dibenzoylmethane)
diglyme	bis(2-methoxylethyl)ether
dme	1,2-dimethoxyethane
DMF	N,N-dimethylformamide
dmpe	1,2-bis(dimethylphosphino)ethane
DMSO	dimethylsulfoxide
dpe	cis-1,2-bis(diphenylphosphino)ethylene
dpmH	2,2,6,6-tetramethyl-3,5-heptanedione (dipivaloylmethane)
dppe	1,2-bis(diphenylphosphino)ethane
dppm	bis(diphenylphosphino)methane
$EDTAH_4$	ethylenediaminetetraacetic acid
en	1,2-diaminoethane (ethylenediamine)
etacH	ethyl 3-oxobutanoate (etyl acetoacetate)
$Et_4 dien$	3,9-diethyl-3,6,9-triazaundecane (tetraethyldiethylenetriamine)
hfacH	1,1,1,5,5,5-hexafluoro-2,4-pentanedione (hexafluoro-acetylacetone)
PCy_3	tricyclohexylphosphine
phen	1,10-phenanthroline
py	pyridine
tfacH	1,1,1-trifluoro-2,4-pentanedione (trifluoroacetyl-acetone)
THF	tetrahydrofurane
TIM	2,3,9,10-tetramethyl-1,4,8,11-tetraazacyclotetra-deca-1,3,8,10-tetraene
β-dik (acac, etc)	monoanion of a β-dicarbonyl compound working as an O,O'-chelating ligand or a counteranion in the outer sphere
β-dik-C (acac-C, etc)	monoanion of a β-dicarbonyl compound bound to a metal atom through a carbon atom
β-dikH	neutral molecule of a β-dicarbonyl compound in general

β-dik-*O* (acac-*O*, etc)   monoanion of a *β*-dicarbonyl compound serving as an O-unidentate ligand

β-dik(2-)   dianion of a *β*-dicarbonyl compound

η^3-acac, η^3-etac   monoanion of acacH or etacH bound to a metal atom through three carbon atoms in a trihapto fashion (η-allylic coordination mode)

7 References

1. Cotton, F. A., Wilkinson, G.: Advanced Inorganic Chemistry, p. 61, New York, Wiley 1980[4]
2. Burmeister, J. L.: Coord. Chem. Rev. *1*, 205 (1966); *3*, 225 (1968)
3. Norbury, A. H., Sinha, A. I. P.: Quart. Rev. *24*, 69 (1970)
4. Norbury, A. H.: Advan. Inorg. Chem. Radiochem. *17*, 231 (1975)
5. Balahura, R. J., Lewis, N. J.: Coord. Chem. Rev. *20*, 109 (1976)
6. Hitchman, M. A., Rowbottom, G. L.: Coord. Chem. Rev. *42*, 55 (1982)
7. Pearson, R. G., Henry, P. M., Bergmann, J. G., Basolo, F.: J. ACS *76*, 5920 (1954)
8. Ghazi-Bajat, H., van Eldik, R., Kelm, H.: Inorg. Chim. Acta: *60*, 81 (1982)
9. Murmann, R. K., Taube, H.: J. ACS *78*, 4886 (1956)
10. Grenthe, I., Nordin, E.: Inorg. Chem. *18*, 1869 (1979)
11. Nakamoto, K.: Infrared and Raman Spectra of Inorganic and Coordination Compounds, p. 220, New York, Wiley 1978[3]
12. Basolo, F., Pearson, R. G.: Mechanism of Inorganic Reactions, p. 291, New York, Wiley, 1967[2]
13. Rindermann, W., van Eldik, R., Kelm, H.: Inorg. Chim. Acta *61*, 173 (1982); *68*, 35 (1983)
14. Basolo, F., Burmeister, J. L., Pöe, A. J.: J. ACS *85*, 1700 (1963)
15. Basolo, F., Baddley, W. H., Weidenbaum, K. J.: J. ACS *88*, 1576 (1966)
16. Mares, M., Palmer, D. A., Kelm, H.: Inorg. Chim. Acta *60*, 123 (1982)
17. Buckingham, D. A., Creaser, J. I., Sargeson, A. M.: Inorg. Chem. *9*, 655 (1970)
18. Orhanovic, M., Sutin, N.: J. ACS *90*, 4286 (1968)
19. Isobe, K., Kai, E., Nakamura, Y., Nishimoto, K., Miwa, T., Kawaguchi, S., Kinoshita, K., Nakatsu, K.: J. ACS *102*, 2475 (1980); Isobe, K., Nakamura, Y., Miwa, T., Kawaguchi, S.: Bull. Chem. Soc. Jpn. *60*, 149 (1987)
20. Bau, R. (ed.): Transition Metal Hydrides (Advances in Chemistry Series, No. 167), Washington, D.C., American Chemical Society 1978
21. Bau, R., Teller, R. G., Kirtley, S. W., Koetzle, T. F.: Acc. Chem. Res. *12*, 176 (1979)
22. Humphries, A. P., Kaesz, H. D.: Progr. Inorg. Chem. *25*, 145 (1979)
23. Teller, R. G., Bau, R.: Structure and Bonding *44*, 1 (1981)
24. Venanzi, L. M.: Coord. Chem. Rev. *43*, 251 (1982)
25. Toogood, G. E., Wallbridge, M. G. H.: Advan. Inorg. Chem. Radiochem. *25*, 267 (1982)
26. Moore, D. S., Robinson, S. D.: Chem. Soc. Rev. *12*, 415 (1983)
27. Hlatky, G. G., Crabtree, R. H.: Coord. Chem. Rev. *65*, 1 (1985)

28. Pearson, R. G.: Chem. Rev., *85*, 41 (1985)
29. Johnson, B. F. G. (ed.): Transition Metal Clusters, New York, Wiley 1980
30. Muetterties, E. L.: J. Organomet. Chem. *200*, 177 (1980)
31. Johnson, B. F. G., Lewis, J.: Advan. Inorg. Chem. Radiochem. *24*, 225 (1981)
32. Ref. (11), p. 304
33. Meek, D. W., Mazanec, T. J.: Acc. Chem. Res. *14*, 266 (1981)
34. Abrahams, S. C., Ginsberg, A. P., Knox, K.: Inorg. Chem. *3*, 558 (1964)
35. Gregson, D., Mason, S. A., Howard, J. A. K., Spencer, J. L., Turner, D. G.: Inorg. Chem. *23*, 4103 (1984)
36. Bau, R., Ho, D. M., Gibbins, S. G.: J. ACS *103*, 4960 (1981)
37. Hart, D. W., Bau, R., Koetzle, T. F.: J. ACS *99*, 7557 (1977)
38. Garlaschelli, L., Khan, S. I., Bau, R., Longoni, G., Koetzle, T. F.: J. ACS *107*, 7212 (1985)
39. Appleton, T. G., Clark, H. C., Manzer, L. E.: Coord. Chem. Rev. *10*, 335 (1973)
40. Griffith, W. P., Mockford, M. J., Skapski, A. C.: J. Chem. Soc., Chem. Commun. 407 (1984)
41. Chan, A. S. C., Shieh, H.-S.: J. Chem. Soc., Chem. Commun. 1379 (1985)
42. Mura, P.: J. ACS *108*, 351 (1986)
43. Roziere, J., Williams, J. M., Stewart, Jr., R. P., Petersen, J. L., Dahl, L. F.: J. ACS *99*, 4497 (1977)
44. Petersen, J. L., Johnson, P. L., O'Connor, J., Dahl, L. F., Williams, J. M.: Inorg. Chem. *17*, 3460 (1978)
45. Petersen, J. L., Brown, R. K., Williams, J. M., McMullan, R. K.: Inorg. Chem. *18*, 3493 (1979)
46. Wilson, R. D., Graham, S. A., Bau, R.: J. Organomet. Chem. *91*, C49 (1975)
47. Hart, D. W., Bau, R., Koetzle, T. F.: Organometallics *4*, 1590 (1985)
48. Petersen, J. L., Masino, A., Stewart, Jr., R. P.: J. Organomet. Chem. *208*, 55 (1981)
49. Darensbourg, M. Y., Atwood, J. L., Burch, Jr., R. R., Hunter, W. E., Walker, N.: J. ACS *101*, 2631 (1979)
50. Darensbourg, M. Y., Atwood, J. L., Hunter, W. E., Burch, Jr., R. R.: J. ACS *102*, 3290 (1980)
51. Darensbourg, M. Y., Mehdawi, R. E., Delord, T. J., Fronczek, F. R., Watkins, S. F.: J. ACS *106*, 2583 (1984)
52. Teller, R. G., Williams, J. M., Koetzle, T. F., Burch, R. R., Gavin, R. M., Muetterties, E. L.: Inorg. Chem. *20*, 1806 (1981)
53. Wei, C.-Y., Marks, M. W., Bau, R., Kirtley, S. W., Bisson, D. E., Henderson, M. E., Koetzle, T. F.: Inorg. Chem. *21*, 2556 (1982)
54. Chiang, M. Y., Bau, R., Minghetti, G., Bandini, A. L., Banditelli, G., Koetzle, T. F.: Inorg. Chem. *23*, 122 (1984)
55. Bianchini, C., Mealli, C., Meli, A., Sabat, M.: J. Chem. Soc., Chem. Commun. 777 (1986)
56. Allison, J. D., Cotton, F. A., Powell, G. L., Walton, R. A.: Inorg. Chem. *23*, 159 (1984)
57. Ginsberg, A. P., Abrahams, S. C., Marsh, P., Ataka, K., Sprinkle, C. R.: J. Chem. Soc., Chem. Commun. 1321 (1984)
58. Bau, R., Carroll, W. E., Teller, R. G., Koetzle, T. F.: J. ACS *99*, 3872 (1977)
59. Hlatky, G. G., Johnson, B. F. G., Lewis, J., Raithby, P. R.: J. Chem. Soc., Dalton Trans. 1277 (1985)

60. Bino, A., Bursten, B. E., Cotton, F. A., Fang, A.: Inorg. Chem. *21*, 3755 (1982)
61. Boorman, P. M., Moynihan, K. J., Patel, V. D., Richardson, J. F.: Inorg. Chem. *24*, 2989 (1985)
62. Lehner, H., Matt, D., Togni, A., Thouvenot, R., Venanzi, L. M., Albinati, A.: Inorg. Chem. *23*, 4254 (1984)
63. Delavaux, B., Chaudret, B., Dahan, F., Poilblanc, R.: Organometallics *4*, 935 (1985)
64. Orpen, A. G., McMullan, R. K.: J. Chem. Soc., Dalton Trans. 463 (1983)
65. Ricci, J. S., Koetzle, T. F., Goodfellow, R. J., Espinet, P., Maitlis, P. M.: Inorg. Chem. *23*, 1828 (1984)
66. Hart, D. W., Teller, R. G., Wei, C.-Y., Bau, R., Longoni, G., Campanella, S., Chini, P., Koetzle, T. F.: J. ACS *103*, 1458 (1981)
67. Jackson, P. F., Johnson, B. F. G., Lewis, J., Raithby, P. R., McPartlin, M., Nelson, W. J. H., Rouse, K. D., Allibon, J., Mason, S. A.: J. Chem. Soc., Chem. Commun. 295 (1980)
68. Eady, C. R., Jackson, P. F., Johnson, B. F. G., Lewis, J., Malatesta, M. C., McPartlin, M., Nelson, W. J. H.: J. Chem. Soc., Dalton Trans. 383 (1980)
69. Tachikawa, M., Muetterties, E. L.: Progr. Inorg. Chem. *28*, 203 (1981)
70. Bradley, J. S.: Advan. Organomet. Chem. *22*, 1 (1983)
71. Gladfelter, W. L.: Advan. Organomet. Chem. *24*, 41 (1985)
72. Blohm, M. L., Gladfelter, W. L.: Organometallics *4*, 45 (1985)
73. Vidal, J. L., Walker, W. E., Schoening, R. C.: Inorg. Chem. *20*, 238 (1981)
74. Vidal, J. L.: Inorg. Chem. *20*, 243 (1981)
75. Ciani, G., Garlaschelli, L., Sironi, A., Martinengo, S.: J. Chem. Soc., Chem. Commun. *563* (1981)
76. Parshall, G. W.: Homogeneous Catalysis, New York, Wiley 1980
77. Collman, J. P., Hegedus, L. S.: Principles and Applications of Organotransition Metal Chemistry, Mill Valley, Calif., University Science Books 1980
78. Masters, C.: Homogeneous Transition-metal Catalysis, New York, Chapman and Hall 1981
79. Keim, W. (ed.): Catalysis in C_1 Chemistry, Dordrecht–Boston–Lancaster, D. Reidel Publishing Company 1983
80. Sheldon, R. A.: Chemicals from Synthesis Gas, Dordrecht–Boston–Lancaster, D. Reidel Publishing Company 1983
81. Morrison, J. D. (ed.): Asymmetric Synthesis, Vol. 5. Chiral Catalysis, Orlando, Florida, Academic Press 1985
82. Milstein, D.: Acc. Chem. Res. *17*, 221 (1984)
83. Pruett, R. L.: Advan. Organomet. Chem. *17*, 1 (1979)
84. Kuhlmann, E. J., Alexander, J. J.: Coord. Chem. Rev. *33*, 195 (1980)
85. Chapter 4 of ref. 80
86. Ugo, R.: Hydroformylation and Carbonylation Reactions, p. 135 in ref. 79
87. James, B. R.: Advan. Organomet. Chem. *17*, 319 (1979)
88. Jardine, F. H.: Progr. Inorg. Chem. *28*, 63 (1981)
89. Halpern, J.: Discuss. Faraday Soc. *46*, 7 (1968)
90. Taqui Khan, M. M., Martell, A. E.: Homogeneous Catalysis by Metal Complexes, Vol. I, Chapter 1, New York–London, Academic Press 1974
91. Brothers, P. J.: Progr. Inorg. Chem. *28*, 1 (1981)
92. Vaska, L.: Acc. Chem. Res. *1*, 335 (1968)
93. Zhou, P., Vitale, A. A., Filippo, Jr., J. S., Saunders, Jr., W. H.: J. ACS *107*, 8049 (1985)

94. Kubas, G. J., Ryan, R. R., Swanson, B. I., Vergamini, P. J., Wasserman, H. J.: J. ACS *106*, 451 (1984)
95. Wasserman, H. J., Kubas, G. J., Ryan, R. R.: J. ACS *108*, 2294 (1986)
96. Morris, R. H., Sawyer, J. F., Shiralian, M., Zubkowski, J. D.: J. ACS *107*, 5581 (1985)
97. Ref. 77, p. 333
98. Osborn, J. A., Jardine, F. H., Young, J. F., Wilkinson, G.: J. Chem. Soc. (A) 1711 (1966)
99. Meakin, P., Jesson, J. P., Tolman, C. A.: J. ACS *94*, 3240 (1972)
100. Arai, H., Halpern, J.: Chem. Commun. 1571 (1971)
101. Halpern, J., Wong, C. S.: J. Chem. Soc., Chem. Commun. 629 (1973)
102. Halpern, J., Okamoto, T., Zakhariev, A.: J. Mol. Cat. *2*, 65 (1977)
103. Ref. 77, p. 338
104. Halpern, J., Riley, D. P., Chan, A. S. C., Pluth, J. J.: J. ACS *99*, 8055 (1977)
105. Chan, A. S. C., Pluth, J. J., Halpern, J.: Inorg. Chim. Acta *37*, L477 (1979)
106. Chan. A. S. C., Halpern, J.: J. ACS *102*, 838 (1980)
107. Čaplar, V., Comisso, G., Šunjić, V.: Synthesis 85 (1981)
108. Knowles, W. S.: Acc. Chem. Res. *16*, 106 (1983)
109. Kagan, H. B.: Chiral Ligands for Asymmetric Catalysis, p. 1 in ref. 81
110. Halpern, J.: Asymmetric Catalytic Hydrogenation; Mechanism and Origin of Enantioselection, p. 41 in ref. 81
111. Fryzuk, M. D., Bosnich, B.: J. ACS *99*, 6262 (1977)
112. Horwitz, C. P., Shriver, D. F.: Advan. Organomet. Chem. *23*, 219 (1984)
113. Colton, R., McCormick, M. J.: Coord. Chem. Rev. *31*, 1 (1980)
114. Longato, B., Martin, B. D., Norton, J. R., Anderson, O. P.: Inorg. Chem. *24*, 1389 (1985)
115. Nelson, N. J., Kime, N. E., Shriver, D. F.: J. ACS *91*, 5173 (1969)
116. Kim, N. E., Nelson, N. J., Shriver, D. F.: Inorg. Chim. Acta *7*, 393 (1973)
117. de Boer, E. J. M., de With, J., Orpen, A. G.: J. Chem. Soc., Chem. Commun. 1666 (1985)
118. Johnson, B. F. G., Lewis, J., McPartlin, M., Pearsall, M.-A., Sironi, A.: J. Chem. Soc., Chem. Commun. 1089 (1984)
119. Colton, R., Commons, C. J., Hoskins, B. F.: J. Chem. Soc., Chem. Commun. 363 (1975)
120. Commons, C. J., Hoskins, B. F.: Aust. J. Chem. *28*, 1663 (1975)
121. Colton, R., Commons, C. J.: Aust. J. Chem. *28*, 1673 (1975)
122. Herrmann, W. A., Biersack, H., Ziegler, M. L., Weidenhammer, K., Siegel, R., Rehder, D.: J. ACS *103*, 1692 (1981)
123. Brun, P., Dawkins, G. M., Green, M., Miles, A. D., Orpen, A. G., Stone, F. G. A.: J. Chem. Soc., Chem. Commun. 926 (1982)
124. Manassero, M., Sansoni, M., Longoni, G.: J. Chem. Soc., Chem. Commun. 919 (1976)
125. Masters, C.: Advan. Organomet. Chem. *17*, 61 (1979)
126. Muetterties, E. L., Stein, J.: Chem. Rev. *79*, 479 (1979)
127. Dry, M. E.: The Fischer-Tropsch Synthesis, in: Catalysis-Science and Technology, Vol. 1, (ed.) Anderson, J. R., Boudart, M., p. 159, Berlin – Heidelberg – New York, Springer, 1981
128. Rofer-DePoorter, C. K.: Chem. Rev. *81*, 447 (1981)
129. Balckborow, J. R., Daroda, R. J., Wilkinson, G.: Coord. Chem. Rev. *43*, 17 (1982)

130. Herrmann, W. A.: Angew. Chem. Int. Ed. Engl. *21*, 117 (1982)
131. Anderson, R. B.: The Fischer-Tropsch Synthesis, Orlando, Florida, Academic Press 1984
132 Araki, M., Ponec, V.: J. Catal. *44*, 439 (1976)
133. Brady III, R. C., Pettit, R.: J. ACS *102*, 6181 (1980); ibid. *103*, 1287 (1981)
134. Holt, E. M., Whitmire, K. H., Shriver, D. F.: J. Organomet. Chem. *213*, 125 (1981)
135. van Buskirk, G., Knobler, C. B., Kaesz, H. D.: Organometallics *4*, 149 (1985)
136. Beno, M. A., Williams, J. M., Tachikawa, M., Muetterties, E. L.: J. ACS *103*, 1485 (1981)
137. e.g. Flank, A. M., Weininger, M., Mortenson, L. E., Cramer, S. P.: J. ACS *108*, 1049 (1986)
138. Chatt, J., da Câmara Pina, G. L. M., Richards, R. L. (ed.): New Trends in the Chemistry of Nitrogen Fixation, London, Academic 1980
139. Chatt, J., Dilworth, J. R., Richards, R. L.: Chem. Rev. *78*, 589 (1978)
140. Henderson, R. A., Leigh, G. J., Pickett, C. J.: Advan. Inorg. Chem. Radiochem. *27*, 197 (1983)
141. Allen, A. D., Senoff, C. W.: Chem. Commun. 621 (1965)
142. Carmona, E., Galindo, A., Poveda, M. L., Rogers, R. D.: Inorg. Chem. *24*, 4033 (1985)
143. Anderson, S. N., Richards, R. L., Hughes, D. L.: J. Chem. Soc., Dalton Trans. 245 (1986)
144. Anderson, S. N., Hughes, D. L., Richards, R. L.: J. Chem. Soc., Dalton Trans. 1591 (1986)
145. Sanner, R. D., Manriquez, J. M., Marsh, R. E., Bercaw, J. E.: J. ACS *98*, 8351 (1976)
146. Churchill, M. R., Li, Y.-J., Theopold, K. H., Schrock, R. R.: Inorg. Chem. *23*, 4472 (1984)
147. Churchill, M. R., Wasserman, H. J.: Inorg. Chem. *20*, 2899 (1981)
148. Churchill, M. R., Wasserman, II. J.: Inorg. Chem. *21*, 218 (1982)
149. (a) Rocklage, S. M., Turner, H. W., Fellmann, J. D., Schrock, R. R.: Organometallics *1*, 703 (1982)
     (b) Rocklage, S. M., Schrock, R. R.: J. ACS *104*, 3077 (1982)
150. Takahashi, T., Kodama, T., Watakabe, A., Uchida, Y., Hidai, M.: J. ACS *105*, 1680 (1983)
151. Klein, H. F., Ellrich, K., Ackermann, K.: J. Chem. Soc., Chem. Commun. 888 (1983)
152. Jeffery, J., Lappert, M. F., Riley, P. I.: J. Organomet. Chem. *181*, 25 (1979)
153. Jonas, K.: Angew. Chem. Int. Ed. Engl. *12*, 997 (1973)
154. Krüger, C., Tsay, Y.-H.: Angew. Chem. Int. Ed. Engl. *12*, 998 (1973)
155. Jonas, K., Brauer, D. J., Krüger, C., Roberts, P. J., Tsay, Y.-H.: J. ACS *98*, 74 (1976)
156. Jonas, K., Krüger, C.: Angew. Chem. Int. Ed. Engl. *19*, 520 (1980)
157. Pez, G. P., Apgar, P., Crissey, R. K.: J. ACS *104*, 482 (1982)
158. Chatt, J., Heath, G. A., Richards, R. L.: J. Chem. Soc., Dalton Trans. 2074 (1974)
159. Heath, G. A., Mason, R., Thomas, K. M.: J. ACS *96*, 259 (1974)
160. Chatt, J., Pearman, A. J., Richards, R. L.: J. Chem. Soc., Dalton Trans. 1520 (1976)

161. Chatt, J., Pearman, A. J., Richards, R. L.: J. Chem. Soc., Dalton Trans. 1852 (1977)
162. Manriquez, J. M., Bercaw, J. E.: J. ACS *96*, 6229 (1974)
163. Manriquez, J. M., Sanner, R. D., Marsh, R. E., Bercaw, J. E.: J. ACS *98*, 3042 (1976)
164. Colquhoun, H. M.: Acc. Chem. Res. *17*, 23 (1984)
165. Wagner, E. L.: J. Chem. Phys. *43*, 2728 (1965)
166. Cotton, F. A., Davison, A., Ilsley, W. H., Trop, H. S.: Inorg. Chem. *18*, 2719 (1979)
167. Nelson, S. M., Esho, F. S., Drew, M. G. B.: J. Chem. Soc., Chem. Commun. 388 (1981)
168. Lindqvist, I., Strandberg, B.: Acta Crystallogr. *10*, 173 (1957)
169. Buckley, R. C., Wardeska, J. G.: Inorg. Chem. *11*, 1723 (1972)
170. Fronczek, F. R., Schaefer, W. P.: Inorg. Chem. *14*, 2066 (1975)
171 Chatt, J., Hart, F. A.: J. Chem. Soc. 1416 (1961)
172. Gregory, U. A., Jarvis, J. A. J., Kilbourn, B. T., Owston, P. G.: J. Chem. Soc. (A) 2770 (1970)
173. Alyea, E. C., Ferguson, G., Restivo, R. J.: J. Chem. Soc., Dalton Trans. 1845 (1977)
174. Cannas, M., Carta, G., Cristini, A., Marongiu, G.: J. Chem. Soc., Dalton Trans. 300 (1976)
175. Kabešová, M., Dunaj-Jurčo, M., Serátor, M., Gažo, J., Garaj, J.: Inorg. Chim. Acta *17*, 161 (1976)
176. Bailey, R. A., Kozak, S. L., Michelsen, T. W., Mills, W. N.: Coord. Chem. Rev. *6*, 407 (1971)
177. Jones, L. H.: J. Chem. Phys. *25*, 1069 (1956)
178. Fultz, W. C., Burmeister, J. L., MacDougall, J. J., Nelson, J. H.: Inorg. Chem. *19*, 1085 (1980)
179. Maroney, M. J., Fey, E. O., Baldwin, D. A., Stenkamp, R. E., Jensen, L. H., Rose, N. J.: Inorg. Chem. *25*, 1409 (1986)
180. Pearson, R. G.: J. ACS *85*, 3533 (1963); J. Chem. Educ. *45*, 581, 643 (1968)
181. Ahrland, S., Chatt, J., Davies, N. R.: Quart. Rev. (London) *12*, 265 (1958)
182. Jørgensen, C. K.: Inorg. Chem. *3*, 1201 (1964)
183. Pearson, R. G.: Inorg. Chem. *12*, 712 (1973)
184. Melpolder, J. B., Burmeister, J. L.: Inorg. Chim. Acta *49*, 115 (1981)
185. Clark, G. R., Palenik, G. J.: Inorg. Chem. *9*, 2754 (1970)
186. MacDougall, J. J., Nelson, J. H., Fultz, W. C., Burmeister, J. L., Holt, E. M., Alcock, N. W.: Inorg. Chim. Acta *63*, 75 (1982)
187. Palenik, G. J., Mathew, M., Steffen, W. L., Beran, G.: J. ACS *97*, 1059 (1975)
188. MacDougall, J. J., Holt, E. M., de Meester, P., Alcock, N. W., Mathey, F., Nelson, J. H.: Inorg. Chem. *19*, 1439 (1980)
189. Nelson, J. H., MacDougall, J. J., Alcock, N. W., Mathey, F.: Inorg. Chem. *21*, 1200 (1982)
190. Burmeister, J. L., Hassel, R. L., Phelan, R. J.: Inorg. Chem. *10*, 2032 (1971)
191. Melpolder, J. B., Burmeister, J. L.: Inorg. Chim. Acta *15*, 91 (1975)
192. Gutterman, D. F., Gray, H. B.: J. ACS *91*, 3105 (1969)
193. Weddle, G. H., Burmeister, J. L., Birnbaum, E. R.: Inorg. Chim. Acta *18*, 59 (1976)
194. Kawaguchi, S.: Coord. Chem. Rev. *70*, 51 (1986)

195. Mehrotra, R. C., Bohra, R., Gaur, D. P.: Metal β-Diketonates and Allied Derivatives, New York, Academic Press 1978
196 Joshi, K. C., Pathak, V. N.: Coord, Chem. Rev. *22*, 37 (1977)
197. e.g. Fay, D. P., Nichols, Jr., A. R., Sutin, N.: Inorg. Chem. *10*, 2096 (1971)
198. Hynes, M. J., O'Shea, M. T.: Inorg. Chim. Acta *73*, 201 (1983)
199. Thompson, D. W., Allred, A. L.: J. Phys. Chem. *75*, 433 (1971)
200. Burdett, J. L., Rogers, M. T.: J. ACS *86*, 2105 (1964)
201. Allen, G., Dwek, R. A.: J. Chem. Soc. (B) 161 (1966)
202. Lintvedt, R. L., Holtzclaw, Jr., H. F.: J. ACS. *88*, 2713 (1966)
203. Harries, H. J., Parry, G., Burgess, J.: Inorg. Chim. Acta *31*, 233 (1978)
204 Camerman, A., Mastropaolo, D., Camerman, N.: J. ACS *105*, 1584 (1983)
205. Iijima, K., Ohnogi, A., Shibata, S.: J. Mol. Struct. *156*, 111 (1987)
206. Brown, R. S., Tse, A., Nakashima, T., Haddon, R. C.: J. ACS *101*, 3157 (1979)
207. Williams, D. E.: Acta Crystallogr. *21*, 340 (1966)
208. Jones, R. D. G.: Acta Crystallogr. Sect. B*32*, 1807 (1976)
209. Power, L. F., Turner, K. E., Moore, F. H.: J. Cryst. Mol. Struct. *5*, 59 (1975)
210. Jones, R. D. G., Power, L. F.: Acta Crystallogr. Sect. B *32*, 1801 (1976)
211. Power, L. F., Turner, K. E., Moore, F. H., Jones, R. D. G.: J. Cryst. Mol. Struct. *5*, 125 (1975)
212. Power, L. F., Jones, R. D. G., Pletcher, J., Sax, M.: J. Chem. Soc., Perkin Trans. II, 1818 (1975)
213. Jones, R. D. G.: Acta Crystallogr. Sect. B *32*, 1224 (1976)
214. Jones, R. D. G.: Acta Crystallogr. Sect. B *32*, 2133 (1976)
215. Jones, R. D. G.: Acta Crystallogr. Sect. B *32*, 301 (1976)
216. Jones, R. D. G.: J. Chem. Soc., Perkin Trans. II, 513 (1976)
217. Larson, M. L., Moore, F. W.: Inorg. Chem. *5*, 801 (1966)
218. Nakamura, Y., Kawaguchi, S.: Chem. Commun. 716 (1968); Nakamura, Y., Gotani, M., Kawaguchi, S.: Bull. Chem. Soc. Jpn. *45*, 457 (1972)
219. Koda, S., Ooi, S., Kuroya, H., Isobe, K., Nakamura, Y., Kawaguchi, S.: Chem. Commun. 1321 (1971)
220. Anzenhofer, K., Hewitt, T. G.: Z. Kristallogr. *134*, 54 (1971)
221. Cramer, R. E., Cramer, S. W., Cramer, K. F., Chudyk, M. A., Seff, K.: Inorg. Chem. *16*, 219 (1977)
222. Nakamura, Y., Isobe, K., Morita, H., Yamazaki, S., Kawaguchi, S.: Inorg. Chem. *11*, 1573 (1972)
223. Koda, S., Ooi, S., Kuroya, H., Nakamura, Y., Kawaguchi, S.: Chem. Commun. 280 (1971)
224. Fredette, M. C., Lock, C. J. L.: Can. J. Chem. *53*, 2481 (1975)
225. Fredette, M. C., Lock, C. J. L.: Can. J. Chem. *51*, 1116 (1973)
226. Mason, R., Robertson, G. B., Pauling, P. J.: J. Chem. Soc. (A) 485 (1969)
227. Allen, G., Lewis, J., Long, R. F., Oldham, C.: Nature (London) *202*, 589 (1964)
228. Gibson, D., Lewis, J., Oldham, C.: J. Chem. Soc. (A) 72 (1967)
229. Behnke, G. T., Nakamoto, K.: Inorg. Chem. *7*, 2030 (1968)
230. Hillis, J., Francis, J., Ori, M., Tsutsui, M.: J. ACS *96*, 4800 (1974)
231. Lingafelter, C.: Coord. Chem. Rev. *1*, 151 (1966)
232. Fackler, Jr., J. P.: Progr. Inorg. Chem. *7*, 361 (1966)
233. Pike, R. M.: Coord. Chem. Rev. *2*, 163 (1967)
234. Shibata, S., Onuma, S., Inoue, H.: Inorg. Chem. *24*, 1723 (1985)

235. Shibata, S., Onuma, S., Iwase, A., Inoue, H.: Inorg. Chim. Acta 25, 33 (1977)
236. Cotton, F. A., Rice, G. W.: Nouv. J. Chim. 1, 301 (1977)
237. Cotton, F. A., Elder, R. C.: Inorg. Chem. 4, 1145 (1965)
238. Bullen, G. J., Mason, R., Pauling, P.: Inorg. Chem. 4, 456 (1965)
239. Hursthouse, M. B., Laffey, M. A., Moore, P. T., New, D. B., Raithby, P. R., Thornton, P.: J. Chem. Soc., Dalton Trans. 307 (1982)
240. Bennett, M. J., Cotton, F. A., Eiss, R.: Acta Crystallogr. Sect. B 24, 904 (1968)
241. Maslen, E. N., Greaney, T. M., Raston, C. L., White, A. H.: J. Chem. Soc., Dalton Trans. 400 (1975)
242. Graddon, D. P.: Coord. Chem. Rev. 4, 1 (1969)
243. (a) Nishikawa, Y., Nakamura, Y., Kawaguchi, S.: Bull. Chem. Soc. Jpn. 45, 155 (1972)
     (b) Koda, S., Ooi, S., Kuroya, H., Nishikawa, Y., Nakamura, Y., Kawaguchi, S.: Inorg. Nucl. Chem. Lett. 8, 89 (1972)
244. Boeyens, J. C. A., Devilliers, J. P. R.: J. Cryst. Mol. Struct. 2, 197 (1972)
245. Swallow, A. G., Truter, M. R.: Proc. Roy. Soc. London, Ser. A 266, 527 (1962)
246. Werner, A.: Ber. 34, 2584 (1901)
247. Figgis, B. N., Lewis, J., Long, R. F., Mason, R., Nyholm, R. S., Pauling, P. J., Robertson, G. B.: Nature (London) 195, 1278 (1962); Mason, R., Robertson, G. B., Pauling, P. J.: J. Chem. Soc. (A) 485 (1969)
248. Lewis, J., Long, R. F., Oldham, C.: J. Chem. Soc. 6740 (1965)
249. Gibson, D.: Coord. Chem. Rev. 4, 225 (1969)
250. Gibson, D., Lewis, J., Oldham, C.: J. Chem. Soc. (A) 1453 (1966)
251. Hartman, F. A., Kilner, M., Wojcicki, A.: Inorg. Chem. 6, 34 (1967); Parker, P. J., Wojcicki, A.: Inorg. Chim. Acta 11, 9 (1974)
252. Gibson, D., Johnson, B. F. G., Lewis, J.: J. Chem. Soc. (A) 367 (1970)
253. Allmann, R., Flatau, K., Musso, H.: Chem. Ber. 105, 3067 (1972)
254. Bennett, M. A., Mitchell, T. R. B.: Inorg. Chem. 15, 2936 (1976)
255. Rigby, W., Lee, H.-B., Bailey, P. M., McCleverty, J. A., Maitlis, P. M.: J. Chem. Soc., Dalton Trans. 387 (1979)
256. Swallow, A. G., Truter, M. R.: Proc. Roy. Soc. London, Ser. A 254, 205 (1960); Hargreaves, R. N., Truter, M. R.: J. Chem. Soc. (A) 2282 (1969)
257. Hazell, A. C., Truter, M. R.: Proc. Roy. Soc. London, Ser. A 254, 218 (1960)
258. Lewis, J., Oldham, C.: J. Chem. Soc. (A) 1456 (1966)
259. Nakamura, Y., Nakamoto, K.: Inorg. Chem. 14, 63 (1975)
260. Baba, S., Ogura, T., Kawaguchi, S.: Bull. Chem. Soc. Jpn. 47, 665 (1974)
261. Horike, M., Kai, Y., Yasuoka, N., Kasai, N.: J. Organomet. Chem. 72, 441 (1974)
262. Kurokawa, T., Miki, K., Tanaka, N., Kasai, N.: Bull. Chem. Soc. Jpn. 55, 45 (1982)
263. Ito, T., Kiriyama, T., Yamamoto, A.: Bull. Chem. Soc. Jpn. 49, 3250 (1976)
264. Ito, T., Kiriyama, T., Nakamura, Y., Yamamoto, A.: Bull. Chem. Soc. Jpn. 49, 3257 (1976)
265. Komiya, S., Kochi, J. K.: J. ACS 99, 3695 (1977)
266. Siedle, A. R., Pignolet, L. H.: Inorg. Chem. 20, 1849 (1981)
267. Okeya, S., Kawaguchi, S.: Inorg. Chem. 16, 1730 (1977)
268. Okeya, S., Kawaguchi, S., Yasuoka, N., Kai, Y., Kasai, N.: Chem. Lett. 53 (1976)

269. Okeya, S., Sazaki, H., Ogita, M., Takemoto, T., Onuki, Y., Nakamura, Y., Mohapatra, B. K., Kawaguchi, S.: Bull. Chem. Soc. Jpn. *54*, 1978 (1981)
270. Okeya, S., Yoshimatsu, H., Nakamura, Y., Kawaguchi, S.: Bull. Chem. Soc. Jpn. *55*, 483 (1982)
271. Okeya, S., Nakamura, Y., Kawaguchi, S.: Bull. Chem. Soc. Jpn. *55*, 1460 (1982)
272. Okeya, S., Nakamura, Y., Kawaguchi, S.: Bull. Chem. Soc. Jpn. *53*, 3396 (1981)
273. Matsumoto, S., Kawaguchi, S.: Bull. Chem. Soc. Jpn. *54*, 1577 (1980)
274. Fish, R. H.: J. ACS *96*, 6664 (1974)
275. Tsuji, J., Takahashi, H.: J. ACS *87*, 3275 (1965)
276. Johnson, B. F. G., Lewis, J., Subramanian, M. S.: J. Chem. Soc. (A) 1993 (1968)
277. Johnson, B. F. G., Keating, T., Lewis, J., Subramanian, M. S., White, D. A.: J. Chem. Soc. (A) 1793 (1969)
278. Kane-Maguire, L. A. P.: J. Chem. Soc. (A) 1602 (1971)
279. Mansfield, C. A., Kane-Maguire, L. A. P.: J. Chem. Soc., Dalton Trans. 2187 (1976)
280. Kurosawa, H.: J. Chem. Soc., Dalton Trans. 939 (1979)
281. Golding, B. T., Harrowfield, J. MacB., Robertson, G. B., Sargeson, A. M., Whimp, P. O.: J. ACS *96*, 3691 (1974)
282. Uchiyama, T., Takagi, K., Matsumoto, K., Ooi, S., Nakamura, Y., Kawaguchi, S.: Bull. Chem. Soc. Jpn. *54*, 1077 (1981)
283. Basato, M., Corain, B., Cofler, M., Veronese, A. C., Zanotti, G.: J. Chem. Soc., Chem. Commun. 1593 (1984)
284. Lile, W. J., Menzies, R. C.: J. Chem. Soc. 1168 (1949)
285. White, D. A.: Synth. Inorg. Metal-org. Chem. *1*, 59, (1971)
286. Ito, T., Yamamoto, A.: Organomet. Chem. *174*, 237 (1979)
287. Sasakura, F., Isobe, K., Kawaguchi, S.: Bull. Chem. Soc. Jpn. *58*, 657 (1985)
288. Siedle, A. R.: J. Organomet. Chem. *208*, 115 (1981)
289. Siedle, A. R., Pignolet, L. H.: Inorg. Chem. *21*, 135 (1982)
290. Siedle, A. R., Newmark, R. A., Kruger, A. A., Pignolet, L. H.: Inorg. Chem. *20*, 3399 (1981)
291. Kotake, S., Sei, T., Miki, K., Kai, Y., Yasuoka, N., Kasai, N.: Bull. Chem. Soc. Jpn. *53*, 10 (1980)
292. Ault, J. L., Harries, H. J., Burgess, J.: Inorg. Chim. Acta *25*, 65 (1977)
293. Jones, J. R., Patel, S. P.: J. ACS *96*, 574 (1974)
294. Sekine, T., Hasegawa, Y., Ihara, N.: J. Inorg. Nucl. Chem. *35*, 3968 (1973)
295. Okeya, S., Nakamura, Y., Kawaguchi, S.: J. Chem. Soc., Chem. Commun. 914 (1977)
296. West, R.: J. ACS *80*, 3246 (1958)
297. Hammond, G. S., Nonhebel, D. C., Wu, C.-H. S.: Inorg. Chem. *2*, 73 (1963); Nonhebel, D. C.: J. Chem. Soc. 738 (1963)
298. Ito, T., Kiriyama, T., Yamamoto, A.: Chem. Lett. 835 (1976)
299. Okeya, S., Egawa, F., Nakamura, Y., Kawaguchi, S.: Inorg. Chim. Acta *30*, L319, (1978)
300. Bailey, N. A., Fenton, D. E., Franklin, M. V., Hall, M.: J. Chem. Soc., Dalton Trans. 984 (1980)
301. von Seyerl, J., Neugebauer, D., Huttner, G., Krüger, C., Tsay, Y.-H.: Chem. Ber. *112*, 3637 (1979)

302. Howe, J. J., Pinnavaia, T. J.: J. ACS 91, 5378 (1969); Pinnavaia, T. J., Collins, W. T., Howe, J. J.: ibid. 92, 4544 (1970); Pinnavaia, T. J., McClarin, J. A.: ibid. 96, 3012 (1974)
303. Matsumoto, S., Kawaguchi, S.: Bull. Chem. Soc. Jpn. 54, 1704 (1981)
304. Ooi, S., Matsushita, T., Nishimoto, K., Okeya, S., Nakamura, Y., Kawaguchi, S.: Bull. Chem. Soc. Jpn. 56, 3297 (1983)
305. Tanaka, H., Isobe, K., Kawaguchi, S., Okeya, S.: Bull. Chem. Soc. Jpn. 57, 1850 (1984)
306. Ref. 12, p. 351
307. Anderson, G. K., Cross, R. J.: Chem. Soc. Rev. 9, 185 (1980)
308. Okeya, S., Miyamoto, T., Ooi, S., Nakamura, Y., Kawaguchi, S.: Bull. Chem. Soc. Jpn. 57, 395 (1984)
309. Siedle, A. R., Newmark, R. A., Pignolet, L. H.: J. ACS 104, 6584 (1982)
310. Siedle, A. R., Newmark, R. A., Pignolet, L. H.: J. ACS 103, 4947 (1981)
311. Tezuka, Y., Ogura, T., Kawaguchi, S.: Bull. Chem. Soc. Jpn. 42, 443 (1969)
312. Yanase, N., Nakamura, Y., Kawaguchi, S.: Inorg. Chem. 19, 1575 (1980)
313. Oda, K., Yasuoka, N., Ueki, T., Kasai, N., Kakudo, M.: Bull. Chem. Soc. Jpn. 43, 362 (1970)
314. Smith, A. E.: Acta Crystallogr. 18, 331 (1965)
315. Kanda, Z., Nakamura, Y., Kawaguchi, S.: Inorg. Chem. 17, 910 (1978)
316. Baba, S., Sobata, T., Ogura, T., Kawaguchi, S.: Bull. Chem. Soc. Jpn. 47, 2792 (1974)
317. Horike, M., Kai, Y., Yasuoka, N., Kasai, N.: J. Organomet. Chem. 86, 269 (1975)
318. Rogers, M. T., Burdett, J. L.: Can. J. Chem. 43, 1516 (1965)
319. Dewar, D. H., Fergusson, J. E., Hentschel, P. R., Wilkins, C. J., Williams, P. P.: J. Chem. Soc. 688 (1964)
320. Raston, C. L., Secomb, R. J., White, A. H.: J. Chem. Soc., Dalton Trans. 2307 (1976)
321. Dewan, J. C., Silver, J.: Acta Crystallogr. Sect. B 33, 1469 (1977)
322. Dewan, J. C., Silver, J.: J. Organomet. Chem. 125, 125 (1977)
323. Dewan, J. C., Silver, J.: Aust. J. Chem. 30, 487 (1977)
324. Dewan, J. C., Silver, J.: J. Chem. Soc., Dalton Trans. 644 (1977)
325. Dewan, J. C., Silver, J.: Acta Crystallogr. Sect. B 33, 2671 (1977)
326. von Seyerl, J., Neugebauer, D., Huttner, G.: Angew. Chem. Int. Ed. Engl. 16, 858 (1977)
327. Imran, A., Kemmitt, R. D. W., Markwick, A. J. W., McKenna, P., Russell, D. R., Sherry, L. J. S.: J. Chem. Soc., Dalton, Trans. 549 (1985)
328. Yanase, N., Nakamura, Y., Kawaguchi, S.: Inorg. Chem. 17, 2874 (1978)
329. Jackman, L. M., Sternhell, S.: Application of Nuclear Magnetic Resonance Spectroscopy in Organic Chemistry, p. 334, London, Pergamon Press 1969[2]
330. Okeya, S., Nakamura, Y., Kawaguchi, S., Hinomoto, T.: Bull. Chem. Soc. Jpn. 55, 477 (1982)
331. Clarke, D. C., Kemmitt, R. D. W., Mazid, M. A., McKenna, P., Russell, D. R., Schilling, M. D., Sherry, L. J. S.: J. Chem. Soc., Dalton Trans. 1993 (1984)
332. Kemmitt, R. D. W., McKenna, P., Russell, D. R., Sherry, L. J. S.: J. Chem. Soc., Dalton Trans. 259 (1985)

333. Okeya, S., Nakamura, Y., Kawaguchi, S., Hinomoto, T.: Inorg. Chem. *20,* 1576 (1981)
334. Okeya, S., Kawakita, Y., Matsumoto, S., Nakamura, Y., Kawaguchi, S., Kanehisa, N., Miki, K., Kasai, N.: Bull. Chem. Soc. Jpn. *55,* 2134 (1982)
335. Otani, Y., Nakamura, Y., Kawaguchi, S., Okeya, S., Hinomoto, T.: Bull. Chem. Soc. Jpn. *55,* 1467 (1982)
336. Okeya, S., Matoba, T., Moriguchi, R., Nakamura, Y., Isobe, K., Hata, K.: 35th National Research Conference on the Coordination Chemistry held on October 10, 1985 at Hiroshima, Japan
337. e. g. Chan, T.-H., Brownbridge, P.: J. Chem. Soc., Chem. Commun. 578 (1979)

Subject Index

acetylacetonate dianion,
  acac(2-) 98–105
acetylacetonate monoanion,
  acac 84–98
acetylacetonate trianion,
  acac(3-) 105
acetylacetone, acacH 77
ambidentate ligand 59
antisymbiotic effect 69, 70, 74
asymmetric hydrogenation 33

basicity of the carbonyl ligand 36
benzoylacetone, bzacH 78, 82
β-dicarbonyl compounds, β-dikH
– central carbon bonding of
  dianions of 98
– central carbon bonding of
  monoanions of 87
– chelation through terminal
  carbons of dianions of 98
– C,O-chelation of dianions of 102
– C,O,O'-bridging of dianions
  of 99
– C,O,O'-bridging of monoanions
  of 87
– coordination modes for
  75–106
– – for dianions 98–105
– – for monoanions 84–98
– – for neutral molecules 79–83
– – for the acetylacetonate
    trianion 105
– dienediolate chelation of dianions
  of 98
– η-allylic coordination of dianions
  of 100
– η-allylic coordination of
  monoanions of 96

– η^3,O,O'-bridging of dianions
  of 104
– keto-enol tautomerism for 75, 98
– O,O'-bridging of monoanions
  of 85
– O,O'-chelation of monoanions
  of 84
– O-unidentate coordination of
  monoanions of 92–96
– outer-sphere coordination of
  monoanions of 91
– structures of enol molecules
  of 75–78
– terminal carbon bonding of
  monoanions of 97
bonding mechanism of CO 2
Brønsted acidity of the hydride
  ligand 22
butterfly-shaped framework of metal
  clusters 42, 44

carbonyl (CO) ligand
– basicity of 36
– bonding mechanism of 2
– cleavage of 43–45
– coordination modes for
  35–43
– – edge-bridging 35, 37
– – M—CO—M' system 37
– – M_2—CO—M' system 37
– – M_3—CO—M' system 38
– – η^2 (side-on) linkage 39–43
– – face-bridging 35, 37, 39
– – semibridging 35
– – terminal (unidentate) bond-
    ing 35, 37, 39
– reduction of 43–45
cationic Rh(I) complexes 31

central carbon bonding of a β-dik monoanion 87–91
chiral diphosphine 33
counterion effect 73

dianions of β-dikH, coordination modes 98–105
diazene complex 55, 56
diazenido ligand 55, 56
dibenzoylmethane (dbmH) 77, 78
dienediolate chelation of β-dik dianions 98
dihydrogen (H$_2$)
– complex 27
– heterolytic splitting of 25, 26
– homogeneous activation of 25–27
– homolytic splitting of 25, 26
dinitrogen (N$_2$) ligand
– coordination modes for 45–54
– – end-on bridging 49–51
– – end-on unidentate coordination 47
– – end-on:side-on μ$_3$-bridging 54
– – side-on bridging 52
– protonation of 55–57
dipivaloylmethanate (dpm) complex 85

edge-bridging of CO 35, 37
edge-bridging of the hydride ligand 18
electronic effects of ancillary ligands 69, 71
end-on bridging of N$_2$ 49–51
end-on:side-on μ$_3$-bridging of N$_2$ 54
end-on unidentate coordination of N$_2$ 47
end-to-end bridging of NCS$^-$ 62–65
η 3
η-allylic coordination of a β-dik dianion 100
η-allylic coordination of a β-dik monoanion 96
η^2(C,C') coordination of a β-dikH molecule 83
ethyl acetoacetate (etacH) 96

face-bridging of CO 35, 37, 39
face-bridging of the hydride ligand 18
Fischer-Tropsch (F.T.) synthesis 39, 43
– carbide-methylene mechanism of 43

hard acid 68
hard base 67
hard metal 69
heterolytic splitting of H$_2$ 25, 26
homogeneous activation of H$_2$ 25–27
homolytic splitting of H$_2$ 25, 26
hexafluoroacetylacetone (hfacH) 77, 78, 95
hydrazido(2-) ligand 55
hydride ligand
– Brønsted acidity of 22
– chemical reactions of 21–34
– coordination modes for 7–21
– – edge-bridging 18
– – face-bridging 18
– – interstitial 20
– – M(μ-H)$_2$M system 15
– – M(μ-H)$_3$M system 16
– – M(μ-H)$_4$M system 17
– – M(μ-H)(μ-X)$_n$M system 18
– – terminal (unidentate) bonding 8–12
– – unsupported M—H—M system 12–14
– trans influence of 11, 31
hydrogenation of olefins 29, 31

insertion reaction 24, 30
interstitial hydrogen atom 20
interstitial light atom 21
intramolecular migratory insertion 24, 32
IR assay of the coordination modes of NCS$^-$ 65
Iron-molybdenum cofactor 45
isotopic exchange 26

Keto-enol tautomerism of acetylacetone 75, 98
keto-enol tautomerism of β-dikH 75

Lewis acid 68
Lewis base 67
ligand, classification 1
linkage isomerization 3–5
linkage isomers 3–5, 62, 92, 97

Metal cluster 7, 18, 20, 21
– butterfly-shaped framework
  of 42, 44
molecules of β-dikH, coordination
  modes 79–83
monoanions of β-dikH, coordination
  modes 84–98
μ 2

nitrito-nitro linkage isomers 5
nitrogenase 45
– iron-molybdenum cofactor of 45
nitrogen fixation 45

olefin hydrogenation 31
O,O′-chelation of a β-dik anion
  84
O,O′-chelation of a β-dikH keto
  molecule 81
O-unidentate coordination of a β-
  dikH enol molecule 82
O-unidentate coordination of a β-
  dik monoanion 92–96
oxidation of H₂ 26
oxidative addition 8, 18, 23, 27, 29,
  31, 32, 33

Principle of Hard and Soft Acids
  and Bases (HSAB) 67
protonation of the coordinated
  CO 44
protonation of the coordinated
  N₂ 55–57

reduction of CO 43–45
reductive elimination 23, 24, 27, 30,
  33

semibridging of CO 35
side-on bridging of N₂ 52
soft acid 68
soft base 67
soft metal 69
solvent effect 72
steric effects of ancillary ligands 70
symbiosis 69
symbol μ 2
symbol η 4

thenoyltrifluoroacetone 77
thiocyanate ligand
– coordination modes for 59–66
– – end-to-end bridging 62–65
– – IR determination 65
– – μ(N) bridging 60
– – μ(S) bridging 61
– – unidentate coordination
  67–74
three-center, two-electron (3c–2e)
  bond 14
trans influence 8, 11, 31, 63, 69, 70
trifluoroacetylacetone 92

unidentate coordination of
  NCS⁻ 67–74
unidentate (end-on) coordination of
  N₂ 47
unidentate (terminal) coordination of
  CO 35, 37, 39

Vaska's complex 27

Wilkinson complex 29, 33